Process Implementation Through 5S

LAYING THE FOUNDATION FOR LEAN

Process Implementation Through 5S

LAYING THE FOUNDATION FOR LEAN

Drew Willis

CRC Press
Taylor & Francis Group
Boca Raton   London   New York

CRC Press is an imprint of the
Taylor & Francis Group, an **informa** business

A PRODUCTIVITY PRESS BOOK

CRC Press
Taylor & Francis Group
6000 Broken Sound Parkway NW, Suite 300
Boca Raton, FL 33487-2742

© 2016 by Taylor & Francis Group, LLC
CRC Press is an imprint of Taylor & Francis Group, an Informa business

No claim to original U.S. Government works

Printed on acid-free paper
Version Date: 20151021

International Standard Book Number-13: 978-1-4987-4715-8 (Paperback)

Library of Congress Cataloging-in-Publication Data

Names: Willis, Drew (Drew D.), author.
Title: Process implementation through 5S : laying the foundations for lean /
author, Drew Willis.
Description: Boca Raton : Taylor & Francis, 2016. | Includes bibliographical
references and index.
Identifiers: LCCN 2015038094 | ISBN 9781498747158
Subjects: LCSH: Manufacturing processes--Standards. | Factory
management--Standards. | Industrial efficiency.
Classification: LCC TS183 .W57 2016 | DDC 670--dc23
LC record available at http://lccn.loc.gov/2015038094

Visit the Taylor & Francis Web site at
http://www.taylorandfrancis.com

and the CRC Press Web site at
http://www.crcpress.com

Contents

SECTION II   Do–Check–Adjust

SECTION III   Sustain

Foreword

Drew Willis, with his hands-on floor experience at Toyota, has captured the essence of "getting started" on your Lean journey in any industry. In working with many companies and leaders in their Lean culture transformations, they frequently ask me, "where do I start? Tell me what steps I need to take?"

This book answers those questions in a very practical way. It's no accident that 5S (Sort, Straighten, Shine, Standardize, and Sustain) and Standardization are the foundation of the Toyota House, as they are the processes on which to build. These "Lean tools" are not an end to a means, but rather a means to an end. The purpose is not to implement these tools, check off the box, and then move to the next area; they are tools designed to engage the value-added "floor members" to set standards in order to identify problems, and then to engage these same members to solve these problems. So many times, I see organizations mistakenly implement these tools without the team understanding the "why" behind them, and how these tools can literally engage the team in continuous improvement and problem solving. I've seen examples when implementing these tools in a "silo," without this purpose of "why" being made clear and without them leading to improvement, has actually made the performance of the team worse. Doing it effectively takes doing the process the right way.

Process Implementation Through 5S explains, in practical detail, the steps of "how to" walk through this process with your team and to "do it right." The "process" it focuses on is Plan–Do–Check–Act (PDCA). Willis uses PDCA to outline the book and explains how 5S and Standardization are more than tools to be implemented, but are a process. This book also explains how 5S and Standardization are not only foundational parts of the PDCA for your Lean transformation, but also how they actually are PDCA processes within themselves. After reading this book, you will understand how to use both of these tools to help you stabilize, which they will, and also help the team to see problems, which then you can solve. This is the continuous improvement culture that we're striving for in Lean, and most organizations and Lean leaders miss this simple, but powerful, point.

This book not only illustrates this point but also gives you the "map" and process steps to walk through this process with your team. As a result, you will not only have laid the foundation for Lean in your organization, you also will have laid the foundation for PDCA and for "building an army of problem solvers." This is what will ultimately add value to your customers and drive the business results that you are looking for.

Michael Hoseus

Lean practitioner and coauthor of the Shingo Award–winning book,
Toyota Culture, The Heart and Soul of the Toyota Way

1

Introduction

OVERVIEW OF THE PROCESS

Companies are eager to implement Lean into their operations quickly. However, before Lean can be implemented, the proper foundation must be laid through the implementation of standardized work and visual controls. The procedure outlined in this book shows how basic Lean principles can be used to implement standardized work and visual controls through Plan–Do–Check–Adjust (PDCA).* After you have followed the PDCA for Process Implementation outlined in this book, you will have the proper foundation for your company to operate with Lean.

This book provides the road map to not only implement new processes, but also shows how this same implementation process can be used to shore up existing processes and improve upon them. This process, which will be referred to throughout the book as "Process Implementation," has been used in multiple types of production, including batching operations where products not only change on a daily basis, but there are also frequent same-day changeovers. This process provides order, when sometimes all you can see is the chaos.

Many companies are kicking off 5S (Sort, Set in Order, Shine, Standardize, and Sustain) initiatives to clean up the work area and marking/taping off where things are to be located. These intentions are important, and their value should not be undersold. However, what is often overlooked in these initiatives is the effect that a properly implemented 5S and standardized work activity can have in each area (e.g., cost, quality, productivity, culture/people).

If you have gone through a 5S project, or are in or about to undertake a 5S project, you undoubtedly have been exposed to the 5Ss: sort, set in

---

* Plan–Do–Check–Adjust is also commonly referred to as Plan–Do–Check–Act.

1

order, shine, standardize, and sustain. Many people associate these 5Ss with cleaning up the work area and sustaining the appearance. Thus, 5S usually becomes taping off where everything goes, maybe doing some painting and cleaning, and calling it a day. Sustaining the 5S then usually involves having workers clean and fix any taping issue during production downtime and at end of shift. Although this may accomplish four of the Ss, it leaves out the most important one, and the one that this book will focus on, standardize. Process Implementation also takes a different approach from a normal 5S by incorporating standardized work into the standardize element, rather than just standardizing the 5S.

Process Implementation begins with the standardize element of 5S by looking at the work processes, then standardizing and improving upon them using 5S principles, rather than implementing the first three 5S elements, and then standardizing those three. Doing this also differs from the regular 5S in that the main focus of this process begins by looking at the steps of the work process, and then uses the other 5S elements to improve upon those standardized work steps, a kaizen, rather than adding 5S elements into current standardized work.

Implementing visual controls into a current process assumes that the process is standardized and efficient. Integrating 5S into current standardized work, or practicing the regular 5S approach, will eventually show gaps to the standard, which can be improved through continual auditing and kaizen efforts. By using 5S elements to improve the process during the 5S activity, or doing this process, gaps to the standard will be identified in the process that will be improved, and the 5S will reflect this improved process. Thus, you find the gaps in the process during the 5S so that you implement the improved process and the 5S for this improved process simultaneously. Figure 1.1 shows the difference between the traditional 5S, implementation approach and the Process Implementation. By starting with the goal of standardization rather than 5S and engaging the employees in improvement opportunities, the implemented 5S will be much more effective and the process more efficient.

By taking on this additional work on the front end, and dedicating extra planning, you will save time and effort, and also have the best process in place to 5S. Designing an area, setting visuals, and purchasing new tools and equipment for processes that will be improved after the 5S would be both costly and wasteful. This process prevents double work. Also, by watching the work, documenting the steps, and analyzing, anything that is not part of the standardized work will become evident. A good

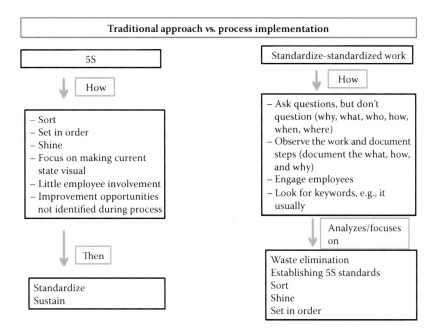

FIGURE 1.1
The process flow on the left shows a traditional 5S implementation, where you start with 5S as the goal, and implement by making the current state visual. The Process Implementation flow (right) explains the process used during Documenting and Analyzing the Work. Starting with the goal of establishing or improving the standard, you will follow the "How" steps. In working through the "How," your focus on what you see and hear should be on Waste Elimination, Sort, Shine, and Set in Order.

example of how this process brings to light issues not seen in traditional 5S involves employee movement. In traditional 5S, the current location for each item would be marked off, regardless of the item's location in relation to the process. Working through the Process Implementation, not only does watching the work reveal the unnecessary movement, but talking to the employees helps establish a more suitable location for items.

5S is not the only focus of Process Implementation, as we will see with the Process Implementation House (Figure 1.2), but it is the 5S that drives how you implement the processes. Process Implementation Through 5S increases efficiency, reduces cost, makes training new employees easier, sets up sustainable systems, and allows you to reach the foundation for operating by using Lean principles, as shown in the Lean House (Figure 1.3).

Lean is an operating strategy that seeks to maximize operational efficiency by creating value for the end customer. Operating with Lean

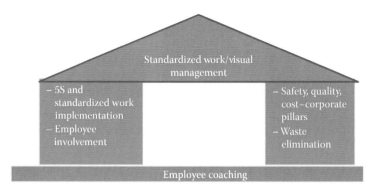

FIGURE 1.2
The Process Implementation House outlines how to achieve stability through standardized work and 5S/visual controls.

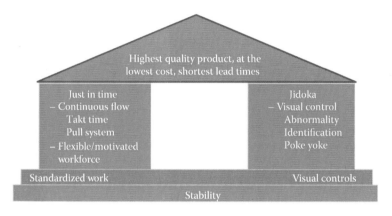

FIGURE 1.3
The Lean House portrays the relationship of the Lean elements as a tiered structure, using a house as a metaphor. Stability lies as the true foundation of the house, followed by a subfoundation that includes standardized work and visual controls. Once this foundation is established, pillar of just-in-time production (jidoka) and employee engagement allow you to reach the top of the house (kaizen), or producing the highest quality product at the lowest cost.

principles is an overall mindset, not focusing on one specific area or process, but an ideal that focuses on attaining the best quality, at the lowest cost, while having high employee morale and the optimal safety. What Lean really is all about is the perpetual identification and elimination of waste from work processes, waste being anything in the process that does not add value. The following are the eight types of waste (with brief definition):

1. **Overproduction**—Producing more than is required.
2. **Inventory**—Generally caused by overproduction, inventory waste occurs when people and costs are incurred for storage of inventory or if it cannot be used and becomes obsolete.
3. **Transportation**—Usually caused by excessive inventory or poor layout; moving items for production should be done quickly (point A to point B) and efficiently (single handling).
4. **Motion**—The primary cause is poor workstation layout. This can be body movements while performing the work or the motion of looking for needed items to perform the work.
5. **Unnecessary Processing**—This can be caused by changes in other processes or just a lack of standard for the process. This involves doing more than is necessary to produce the product.
6. **Defects/Quality**—Again, lack of standardized work is a major cause of defects and quality issues. The results of defects are costs associated with reinspecting, reworking, and inventory waste (see item 2).
7. **Waiting**—Waiting in any form is waste: waiting for machine, waiting for operator, waiting on transportation, waiting because of quality issues.
8. **Underutilizing People***—Losing time, ideas, skills, improvements, and learning opportunities by not engaging or listening to your employees.

Throughout this book, you will see examples of how the Process Implementation addresses certain types of waste, as well as specific examples that prove out the Process Implementation.

Once you are operating with Lean principles, you are beyond Process Implementation but you are also looking to drive out waste. As will be discussed throughout this book, Process Implementation sets the proper foundation for being able to operate Lean by creating standardized work and visual controls. Once you build this foundation, you will operate on a slightly different set of principles, which are a higher level than those used during the implementation (see Figure 1.3).

Think of Process Implementation as somewhat of a "Starter House." Process Implementation is simple but has all the basic amenities of the Lean House; through time, you can upgrade to a larger and more robust house.

---

* There were initially only seven types of waste; Jeff Liker introduced the eighth type of waste in *Toyota Way*, 2004.

Even after you move onto the Lean House, the Process Implementation can be useful if you are doing a kaizen (as this process teaches how to improve a process), changing the area because of a model change, or implementing a new process because of a reaction to a problem in the process. If you feel that your company has a strong Lean foundation, do an assessment, and if you don't like the results, the process outlined in this book will guide you through the process of shoring up your foundation.

The below explains the different elements of the Process Implementation House shown in Figure 1.2.

Top Floor—Fully Implemented Processes/Standardized work/Visual Management

Pillars

1. 5S and Standardized Work Implementation—Just putting in standardized work won't set the proper foundation for Lean, but efficient processes with visual controls that have removed waste provides the proper foundation.

2. Employee involvement—This pillar will be seen throughout each step of the Process Implementation, as engaging the employees, getting their feedback, and getting them to take ownership of the area will improve morale and spur employees to drive improvement.

3. Corporate pillars—Safety, quality, and cost—Consider apparent and potential safety and quality issues for all processes. For each step in the process, look at items that are contributing to safety or quality issues. Also, look at how the step could have safety or quality implications. Focus on cost implications as well, with the goal being low cost or no cost. The focus on these pillars during the Implementation is about driving out waste.

Foundation—Employee coaching—Coaching is the foundation because you must coach why the process is being improved at each step in the PDCA. Employees must understand why the implementation is important so that they can apply the principles they learned during the Process Implementation to sustain their area and drive improvement once you are operating Lean.

If you read from the bottom of the Process Implementation House up through the Lean House, you can see that the top floor of the Process Implementation House matches the foundation for the Lean House. What is missing from the top floor of the Process Implementation House is kaizen; however, implementing processes is itself a kaizen activity, so as you are implementing processes to attain standardized work with visual controls, you are already practicing a Lean principle, kaizen.

Figure 1.4 shows the flow of the Process Implementation that this book walks you through. As evidenced by the title of this book, the majority of this book focuses on the initial action of implementing standardized work and visual controls. However, it also goes into detail in completing the following necessary steps: training the standardized work, workers performing to the standard, auditing the standardized work and 5S, and taking action when the results are out of standard. Countermeasures are covered in detail in the auditing section, as anything that is out of standard needs to have a countermeasure to fix the issue. The implementation will not be sustained unless accountability and proper systems are put in place to ensure sustainment and improvement. The full cycle of PDCA must be followed, as planning and doing set the proper foundation that must be maintained through auditing (checking) and making proper adjustments. The importance of sustaining is evidenced in Figure 1.4 by the amount of items that are included in the Check and Adjust areas.

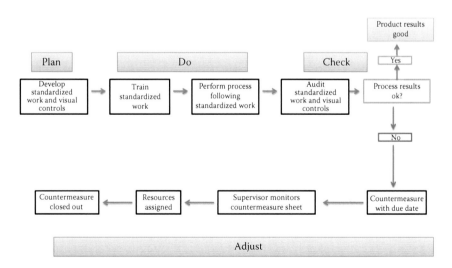

FIGURE 1.4

Standardized Work and 5S Work Flow show the process flow that the Process Implementation will follow. Although each box may have its own PDCA cycle, the overall process is one large PDCA cycle.

5S OVERVIEW

5S is best practiced for arranging and maintaining proper organization of the workplace. 5S offers the following benefits:

1. Improves employee safety and morale—The Occupational Safety and Health Administration reports that lost productivity stemming from injuries and illnesses costs companies $60 billion per year. Also, people like to work in an organized and clean environment; this creates a sense of pride in their work.
2. Reduces on nonvalue-added "searching" by employees—Employees spend less time going and getting the necessary items.
3. Improves utilization of area—The floor space required for a process is decreased when 5S is implemented.
4. Improves quality and reduces mistakes—The risk of mistake is less when employees are not rushed, and only needed items are located in the work area and have a designated location.
5. Improves Culture—Employees take greater pride in their work, and it fosters enthusiasm from workers individually, and teamwork from the group.

Not only do the visual controls that are implemented increase efficiency, but organization reduces defects and safety incidents. If you have ever toured a Lean company such as Toyota or Honda, you were undoubtedly impressed with its neatness and cleanliness. These same Lean companies will tell you that 25–30% of all quality defects are directly related to safety, order, and cleanliness of the workplace (Brennan, 2011, p. 128). 5S consists of five different elements, all starting with the letter "S" that must be followed. These steps are briefly explained in the following list.

1. *Sort*—Only needed items are in the work area.
   Overview of Step: In sort, you will go through the work area and determine what items are needed and what items are not needed.
2. *Set in order*—"A place for everything, and everything in its place."
   Overview of Step: Determine the location and quantity of all needed items. The location and function of the items should be intuitive for someone walking through the work area.

3. *Shine*—Perform initial cleaning and set up cleaning as a daily activity.

   Overview of Step: Don't make cleaning a monotonous task, but invite operators to share what items they notice on their daily cleaning. Keeping a clean workplace not only keeps it safe, but it can expose issues that are not easily noticed otherwise.

4. *Standardize*—In typical 5S, the standardize element is merely maintaining and monitoring the first 3Ss.

   Overview of Step: This will be discussed in great detail throughout the book, and involves not only standardizing the first 3Ss, but combining 5S standards into standardized work to make it more efficient. The 5S standard does not become part of the standardized work document, but creates visual identification items that make up the standardized work (e.g., min/max of material, location of tool, defined workspace).

5. *Sustain*—Train employees and audit the area for 5S and standardized work compliance.

   Overview of Step: Create systems to check the 5S standards. This can be accomplished through audit check sheets that can be used by various levels of management to verify steps of processes followed; maps of the area may also be used to ensure 5S is being followed to standard.

Of these five steps, Standardize and Sustain are the most difficult to achieve, and thus are the two that a majority of this book will focus on. However, sort, set in order, and shine are vital to the success of the process or processes in an area and must not only be implemented effectively, but are also the items most easily identified that show a process is out of standard or that no standard exists when going to the Gemba (Japanese word meaning the place where the work is performed) and observing the work.

Simply organizing everything is not enough to make 5S effective, but adequate visual controls must be implemented to identify item locations and provide a visual aid to determine if something is out of standard. Everything in the work area should be 5Sed, and any new items that are incorporated into a 5S should be fully implemented into the process (i.e., adding to standardized work, adding to 5S checklist). Later in this book, we will discuss the obstacle of getting employee buy-in for labeling everything because of the employees' feeling that, "…well everyone knows what that is, so there is no need to label it." I like to think of 5S as taking a road trip. Unless you have been to the exact place that you are going, you will need to adequately plan by getting the address. The people living in the area undoubtedly could get to this place without directions or needing to look at road signs, but whether you asked locals for directions or used Global Positioning System (GPS), signs of all sorts will be your guide. These signs will not only get you to your destination in the form of road names, highway, interstate signs, etc., but they will also state the standard for driving on different roads in the area, such as speed limit, yellow lines, stop signs, and stop lights. Just as outsiders to an area need signs to guide them, 5S is important so that outsiders and new employees can immediately find "their way," find what they need, or easily determine the standard. I have heard it said that after an area has gone through 5S, you should be able to take an elementary school child through, and they should be able to understand exactly what everything is and its purpose.

STANDARDIZED WORK OVERVIEW

As this 5S process is really all about implementing 5S through Standardized Work, standardized work will also be explained at length in this book, because both 5S and standardized work are necessary to establish the stability for Lean; ensuring that both are correctly implemented is vital.

Standardized work is the method to record the best practice for performing a process; it is the primary tool for providing the work method and defines what, how, and why the work is to be performed (in addition to who and where; however, the what, how, and why are the elements that need to be taught). It is a written and agreed-upon method that ensures that a process is safe, consistent, and efficient. When standardized work is implemented, the standard becomes the best way that is known to perform the process, and must be followed by everyone while it is the standard. Naturally, the standard can always be improved upon, and management should encourage improvement, but while there is a standard in place, it must be followed.

Standardized work and what companies often refer to as "standards" are different. Procedures that are considered standard refer to operating rules, such as defining 5S requirements; color coding of area, items, tools, etc.; work in process (WIP); inventory levels; and material flow. Employees are trained on the standards and are required to maintain the standard, but these standards are not required to be included in the standardized work, as updating them would prove too time consuming. Making them visual and training employees on them will suffice. We will see later an example of an employee identifying that their broom station was out of standard, as it was easily identified that the place marked broom was empty (Figure 9.4). However, the important distinction to make is that although you are implementing 5S through standardized work, the 5S that is being implemented is a standard to follow, which makes the standardized work more efficient.

Having standardized work that is followed by all employees on all shifts has numerous benefits. Those benefits include:

1. Stability—Having standardized work provides the stability needed to have Lean systems by having repeatable processes.
2. Less stressful employees—Employees who know exactly what is required of them will have less stress and perform better.
3. Employee involvement and empowerment—Workers pursuing higher quality through a thorough understanding of the purpose of the process.
4. Consistency in the process—If the process is performed the same each time, the results are consistent.
5. Clarifies the Process—No more "we just know, or we usually"; standardized work defines the process and eliminates ambiguity.
6. Records the Best Practice—Documents the current best practice that is to be followed by all.
7. Assists in training—Standardized work provides a training document that can be used for all new employees, showing them the step-by-step process of completing a certain task.
8. Sets the baseline for improvement—Standardized work is the best way that we know to perform a process "today," but it must be established in order for improvement to occur. You must have a starting point in order to measure improvement.
9. Safety—When the process is designed with employee safety at the forefront, and when everyone knows exactly what to do in each situation, or knows when to raise a hand to question, workplace injuries naturally decrease.

One of the benefits of effective standardized work is that it improves every aspect of your company. Quality is higher, costs are lower, your people are happier and are driving improvement, and the workplace becomes safer. These benefits will be covered throughout the book, but as you can see, having standardized work in place creates not only lower stressed employees and fewer turnover, but also increases the bottom line for your company.

Standardized work and 5S are the foundation for Lean, meaning that without them, attempting to implement other Lean tenets would be ineffective. Stopping to fix the problem (Jidoka), employee involvement, Just in Time (JIT), and Kaizen (continuous improvement) cannot be fully realized without consistent processes, which are made possible through standardized work. The Lean House shown in Figure 1.3 clarifies the interrelatedness of standardized work to the other Lean principles just mentioned.

The following discussion details the different Lean elements shown in Figure 1.3.

Top Floor—Best quality, lowest cost, shortest lead time, best safety, and high morale.
Pillars
1. JIT—Getting only the amount needed of an item when it is needed.
   a. Having continuous flow of materials
   b. Performing a process within the prescribed time, within the takt time
   c. Pull system—the next process pulls from the previous process
2. Highly motivated people—In Lean, the employees drive improvement, and therefore, a highly motivated workforce is essential to get to the top of the house. This is achieved through employee coaching and involvement in the Process Implementation.
3. Jidoka—Stopping to solve problems.
   a. Asking questions to get to the root cause
   b. Poke-yoke—error-proofing the station
Foundation—Stability-standardized work, visual management, and kaizen.
1. This foundation enables you to use the pillars to reach the peak, the top floor. The Implementation Process outlined in this book establishes this foundation.

To build the best Lean House, you first need stability. Once stability is established, standardized work and visual controls become the foundation of the house. Thus, it is important if you are just beginning on your Lean journey to implement standardized work and visual controls at the outset, and if you already have some form of these elements, to work to sustain and improve upon your foundation. Simply documenting the best way to perform a process is only one part of standardized work implementation, but this book walks you through the steps needed to gain stability, and then to establish, document, implement, train, sustain, and improve standardized work. Each step along the journey will require the same overall approach, and although it follows the PDCA cycle, it also explains the reasons for each element by providing the reader with insight into how people think as well as practical examples, and therefore why this process yields great results.

When people think about standardized work, they often cite the efficiency and reduced costs associated with implementation of consistent processes. What is often overlooked is the reduced stress and employee engagement that can be accomplished through a proper implementation and sustainment process. Standardization is sometimes thought of as turning workers into mindless drones. However, just the opposite is true, as standardization allows workers to utilize their creativity and gives them a baseline against which to measure their improvement ideas. Taiichi Ohno, widely considered the father of the Toyota Production System, said that without standards, there can be no kaizen.

BACKGROUND

Workers have a certain skepticism when someone enters their area with the goal of implementing standards or the promise of improvement. Not only has someone entered their space, but these activities generally mean a further invasion of their privacy and the thought of either additional work, higher accountability, or even a potential layoff. Historically, companies have implemented coercive bureaucracy atmospheres, under the guise of Frederick Taylor's teachings. Frederick Taylor, often considered the father of scientific management, developed a scientific method for determining the best way to perform a job, and determined productivity numbers based on the physically possible limits of a job (Adler, 1992). Although generally applied under a coercive bureaucracy, wherein management enforced standardization, adoption of best implements and working conditions, and cooperation,* Taylorism did have some success in increasing productivity. However, the environment that can be created by an enforcement bureaucracy can be a high level of alienation among the organizations' employees.† The technical efficiency of time and motion analysis is often acknowledged when discussing Taylorism, but the dehumanizing effect of the resulting job designs is widely decried.‡

Inversely, a bureaucracy that instills discipline rather than enforcement is not alienating, and can serve as a source of motivation. Increasing formalization of procedures and structures tends to reduce role conflict and ambiguity, thereby increasing work satisfaction and reducing alienation and stress.§ Stress can be caused by not being able to process information as fast as it arrives; stress can also be caused by concern about not having all of the relevant information needed for a task or project (Brenna, 2011, p. 124). As far as standardization of tasks, establishing routines, simplifying processes, and avoiding interruptions can minimize the feeling of being overloaded (Brenna, 2011, p. 126). Formalization is shown to reduce stress, and thereby increase overall work satisfaction; work satisfaction leads to less turnover, a happier workforce, and a higher quality product.

The theories espoused by Adler based on his research of the NUMMI (New United Motor Manufacturing, Inc.) plant, Toyota's initial efforts to understand the American worker and implement Lean, still hold true today. This book explains not only how to properly implement standardized work, but how to do so in a manner that empowers employees to drive improvement. The goal of any company is to eliminate waste and to operate at maximum efficiency. As will be explained in this book, having standardized work systems and 5S as well as the right approach will drive these results.

CULTURE

Culture is how people are incentivized to behave and the way people think, talk, work, and act every day. Corporate culture is based on a philosophy and supported by a management system and structures that allow the

---

* Adler (1992), citing Frederick Taylor's, *The Principles of Scientific Management.*
† Adler (1992) citing Katz and Kahn (1966, p. 222).
‡ Adler (1992) citing Katz and Kahn (1966).
§ Adler (1992), citing multiple sources.

desired behaviors to take place consistently (Zarbo, 2012). When quality is the driving force of culture, it will increase efficiency and productivity, decrease costs, and, in turn, allow the company to achieve lower prices, attract a higher market share, increase profits, and improve customer satisfaction (Zarbo, 2012). These ideals should be the goal of any organization although success should not just be defined by the above metrics alone, but by an engaged workforce perpetually driving process improvements and continually striving for higher targets of quality (Zarbo, 2012).

Having sustainable standardized work is more than Standardized worksheets; although these are integral to the process, no matter how accurate they explain the steps of the process, or how many pictures they have, it means nothing if the workers are not following the standard, if standardized work is not being audited, and if the workers are not working to improve the process. To thrive in each of these areas, the proper bureaucracy—or rather, the structure of the organization—must be established, and the relationship between worker and management must be such that there is mutual respect and trust, or put differently, that the culture of the workplace encourages rather than inhibits employee engagement.

Under the technical function of bureaucracy, the assumption is that work can be fulfilling, and that organization can be experienced as a cooperative endeavor. If employees see at least some overlap between their goals and those of the organization, they might welcome the potential contribution of formalization to efficiency. Under these assumptions, employees will embrace formal work procedures that are appropriately designed and implemented. Well-designed procedures would facilitate task performance and thus augment employees' pride of workmanship.[*]

Having a proper culture means that employees are doing things right the first time, with success being measured by a leader being able to walk away and empowered employees sustaining themselves in pursing higher quality targets by implementing continuous process improvements (Godwyn and Gittel, 2012). This philosophy must be supported by an appropriate management system that empowers the workforce to pursue higher targets of quality while identifying defects blamelessly and then effectively improving the process (Godwyn and Gittel, 2012). Establishing a strong culture can be difficult but is vital to success; significant culture change remains the top challenge in more than 80% of companies, according to a 2005 Aberdeen Group Survey.

---

[*] Godwyn and Gittell (2012) citing Deming (1986) (emphasis added).

When there is a serious error or performance is below the expected level, then traditional Western management goes into action to place the blame, with the assumption in Western culture being that problems are caused by people (Liker and Hoseus, 2008). The problem with this thought process not only creates an atmosphere of distrust, as everyone is pointing their finger at everyone else in an effort to save their own skin, but it also inhibits true improvement by failing to ask the five whys to determine the root cause. Remember back to the Lean House, resting right above standardized work were Jidoka (stopping to fix the problem) and employee involvement, which make up two of the three pillars of the house. Approaching problems with the mentality of finding the "person" at fault rather than the "problem" defies the structure of Lean, which is evident with the Lean House picture, as you cannot build the top level (Highest Quality product at the lowest cost) without the necessary pillars, and these pillars are not strong if problems are treated as a "they" (meaning someone else is at fault), rather than a "we" (meaning everyone working together to solve the issue).

In thinking about this in terms of PDCA, traditional Western culture would "Check" a problem and then place blame. Often, supervisor-level management and above are given their group of responsibilities (e.g., make sure your labor is "good") and left to determine the best way to oversee those responsibilities. Guidance on exact roles and responsibilities is typically unclear, and little—if any—training is given to provide instruction on the most effective way to execute the role. In essence, new people to a position are not given enough of the information they need on day one, but acquire this through process and error over their experience with different scenarios; the better structure would be to provide a detailed description of the roles and responsibilities, as well as specific guidance on how to handle different scenarios, or at least a robust set of guidelines under which to operate (which would be the proper means to "Plan" to enable front line leaders to "Do"). Think about it this way, if you were to ask a frontline supervisor or manager, "Did you do X," and got the response, "No, I didn't think to do X," then that person: (1) had never been told/trained to do X, (2) was not given the guidelines needed to know that X should have been done, or (3) was trained to do X and was given the guidelines needed in order to know how to do X but failed to do it. The first two can be fixed easily, but could have also been provided to the person prior to the problem occurring (i.e., proper "Planning" could have solved the issue). In the third scenario, wherein the person failed to follow standardized work, the person should be retrained at least once on the matter, and proper discipline should be applied for any follow-up instances. This blame mentality is a "you" should have done this, rather than a "why" did this happen, and thwarts necessary action to find the true root cause.

Although companies repeatedly "Check" a problem and place blame, likely without having done the proper planning, to follow PDCA correctly, when a problem occurs with the current process, management should and must determine if there is a process in place and whether that process is effective. If it is not, then the PDCA cycle must be repeated to determine the root cause and implement countermeasures (the necessary change required to resolve the root cause). The root cause could, in fact, be that the worker did not perform the process correctly, but there could be a legitimate reason for the worker's error, such as inadequate or lack of training, management's assumption that the worker knew what they should be doing, or the worker experimenting with an improvement. Finding these reasons will only come about through root cause analysis, and if there is a standard and it was followed, then it must be modified and the employees must be trained to the new standard. If the standard was not followed, then coaching must be done with the employee to ensure that there is no ambiguity about the requirements of the task (we will discuss employee training and proper discipline in Chapter 11). Simply placing blame prevents finding the actual root cause (Eight-Step Problem Solving, a Lean exercise used to determine root cause, is outside the scope of this book, but is the next needed element to implement after the Process Implementation).

The culture of finger pointing only leads to distrust and hinders improvement. Employees are generally resistant to change, but resistance to method changes can be overcome by getting the people involved in the change to participate in making it (Lawrence, 1969). Not only is employee engagement in this process vital, as employee involvement is a pillar of the Lean House, but the employees performing the job are better equipped to know how to improve it because they are intimately involved in it for hours each day.

Building the proper culture of mutual trust is vital in successfully implementing standardized work. Having a blame-free culture where problems are appropriately addressed through root cause analysis rather than the blame game empowers employees to drive improvement, and allows for effective auditing. Real participation is based on respect, and respect is not acquired by just trying; it is acquired when the management realizes it needs the contributions of the shop floor workers (Lawrence, 1969). When employees trust their supervisor, employees are much more likely to have acceptance when change occurs in the workplace (Marksberry, 2013). Trust between the supervisor and subordinate is based on three agreements:

1. All promises must be kept, if a supervisor miscommunicates the outcome of change in the workplace, employees will be less likely to believe their supervisor the next time change is attempted, no matter how appealing (Marksberry, 2013, p. 200). Walking the talk is requisite for any communication effectiveness; communication that is incongruent with behavior can undermine employee motivation and be perceived as manipulation or insincere.*
2. Mutually agreed to items must be put into practice immediately and must go forward as mutually agreed.
3. Supervisors must treat everyone fairly and equally at all times under all circumstances. Although each employee may react differently to change, each supervisor must maintain a fair and consistent approach (Marksberry, 2013, p. 200).

In addition to these necessary agreements, managers must also explain why decisions are made and the implications of these decisions. Organizations with high trust cultures exhibit the following communication practices:

1. Managers explain why decisions are made.
2. Communication occurs in a timely manner.
3. Important information flows continuously.
4. Direct supervisors and other leaders explain the specific implications of environmental and organizational changes to each level of workers.
5. Employee responses to leader communications are validated (Mayfield and Mayfield, 2002, p. 90).

Leader communication is the bridge that transmits behavioral intent to employees, thus creating the foundation for trust (Mayfield and Mayfield,

---

* Mayfield and Mayfield (2002), citing Dulek and Fielden (1990) and Goffman (1959).

2002). All of these practices are necessary to build the proper trust needed for implementation and employee engagement for improvement, and these will be further expounded upon in later chapters; the process of communicating during the implementation and auditing process is directly correlated to the success of the overall process. This is why you will see such high employee involvement in the Process Implementation. Adler (1992, p. 65) points out that you cannot improve a process you don't understand, and explaining the "why" gives the employee the necessary understanding to be able to make improvements. As far as the causal effect between culture and improvement, Adler points out that the key cultural prerequisite for kaizen is a climate in which the appearance of incidents/problems is welcomed as an opportunity for learning, rather than a sign of failure to be hidden from view (Adler, 1992, p. 67). Explaining the "why," which is done during job instruction, also enables a no-blame culture to exist, as a successful Job Instruction program results in having a no-blame culture (Huntzinger, 2006).

In addition to building trust and explaining the potential waste elimination and cost benefits of the Process Implementation, it is important to stress individual benefits, as workers will ask "what's in this for me?" In essence, improving processes will make the work easier for the employee, as the Process Implementation seeks to put in the most efficient and safest process, which gives the employee all the tools needed to perform the task successfully. Also, the benefits of a less stressful organized workspace and the fact that they are learning a highly transferrable skill set that can be used in any environment are key benefits. This sell is for the possibility of learning valuable skills for possible internal promotions, but it also makes the employees marketable elsewhere if they have opportunities for advancement outside of your company.

They may ask whether they will receive raises for these new skills, which in the majority of instances they will not. Remember as a spokesman for the company, you should deflect any negative comments made by the workers, and emphasize that better quality and productivity will make the company more money, which they will see benefits from (being employed, raises, bonuses, etc.). Employees don't like change, but change is perpetual, and just because the way in which they are performing their everyday tasks changes does not mean that they will be receiving higher wages. At some point, some employees may need to be told that this is their job and they have to do this, which should be used as a last resort. Generally, employees will see the benefits and buy into the process without this type of questioning, but you should expect one or two employees to ask these types of questions.

Developing your people properly is vital to the success of Process Implementation, and whether those people stay or whether they seek a new challenge elsewhere, they need to be properly developed while employed at your company. It is important to remember that changing a culture cannot be done overnight, and will be much more difficult to instill into everyone than the standardization of processes. Implementing the proper culture could take years to accomplish, but rather than lose heart because of the duration, continue to follow the same approach and abide by the above communication principles. Changing the overall culture might prove slow, but following the guidelines to change culture will begin to show results immediately.

In summary,

- Culture must be built on trust.
- Trust empowers employees.
- Empowered employees embrace formal work procedures.
- Well-designed procedures create employee pride and drive improvement.

FORMAT OF THIS BOOK

This book will educate you on how to implement processes into any business or manufacturing facility with the goal of fully implementing standardized work and visual controls. Figure 1.5 shows the PDCA flow that you will be using throughout your Process Implementation project. The importance of PDCA will be outlined in the following chapters, but the most critical element of PDCA is that you plan for every possible contingency surrounding your processes before you put in a pilot (the Do phase, which, after checking and adjusting, may put you back into the Plan phase). Often, companies want to implement once they have a basic plan in place, without thinking through all the variables. This seems to be basic human nature, as in our personal lives we are often taking action once we have an initial need for something. I often run to the grocery store as soon as a couple things that I need come to mind. As soon as I get home, I'm usually asked by my wife if I thought to get a number of other things, to which I usually reply, no. You see, I don't really plan through my grocery

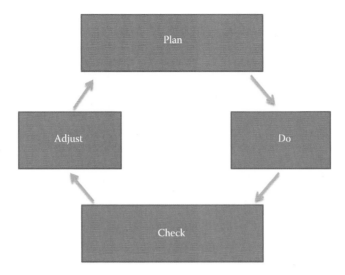

FIGURE 1.5

The flow of Plan–Do–Check–Adjust, a cyclical cycle that is perpetual in Lean. It is the way in which you should approach making any decision. Adjust is sometimes referred to as "Act," but because you are in essence making adjustments based on your "Check," using "Adjust" makes more sense.

trip outside of items that I know I need; if I planned before I actually left for the grocery store, I would be much more efficient and not continue in this cycle.

Although it will be referred to as PDCA throughout the book, it should be written as PDCA, with the Plan being a larger element than the other three, as at least half of your time should be devoted to planning.

REFERENCES

Adler, Paul S., *The "Learning Bureaucracy": New United Motor Manufacturing, Inc.*, School of Business Administration, University of Southern California, p. 59, 1992.

Brennan, Linda L., *The Scientific Management of Information Overload*, Mercer University Journal of Business and Management, Vol. 17, No. 1, p. 128, 2011.

Godwyn, Mary and Jody Hoffer Gittell, *Sociology of Organizations: Structures and Relationships*, SAGE Publications, Inc., Thousand Oaks, CA, p. 109, 2012.

Huntzinger, Jim, *Why Standard Work Is Not Standard: Training within Industry Provides an Answer*, Target Vol. 22, No. 4, p. 11, 2006.

Lawrence, Paul, *How to Deal with Resistance to Change*, Harvard Business Review, January 1969.

Liker, Jeffrey and Michael Hoseus, *Toyota Culture, The Heart and Soul of the Toyota Way*, McGraw-Hill, New York, p. 165, 2008.

Marksberry, Phillip, *The Modern Theory of the Toyota Production System, A Systems Inquiry of the World's Most Emulated and Profitable Management System*, CRC Press, Taylor & Francis Group, Boca Raton, FL, p. 199 (2013).

Mayfield, Jacqueline and Milton Mayfield, *Leader Communication Strategies; Critical Paths to Improving Employee Commitment*, American Business Review, p. 90, June 2002.

Taylor, Frederick, *The Principles of Scientific Management*, Harper and Brothers, New York, 1911, p. 83.

Zarbo, Richard J., *Creating and Sustaining a Lean Culture of Continuous Improvement*, American Journal of Clinical Pathology, p. 321, 2012.

Section I

Plan

As illustrated in the Lean House in Figure 1.4, underneath standardized work and visual controls is stability. This means that, even before you rush into implementation, you must determine whether the processes have the requisite stability needed for implementation. The following discussion explains how to determine the stability of process, how to effectively gain the necessary stability, and how to provide ancillary items that, if properly implemented, will drive the implementation and, down the road, improvement.

2

Identify the Area

The first step in Process Implementation is to identify the area, or even a single process, on which to focus your efforts. Depending on the type of resources that are at your disposal for the implementation, it is better to start small and focus on one area at a time rather than taking on an entire company or plant (similar to implementing a pilot). You can either be implementing a process because you are reacting to something on the floor, or you can be doing an area that needs strong standardized work and visual controls.

When reacting to an issue, you are addressing visible issues. Visible quality issues are those involving rejects, rework, scrap, and inspection. When any of these are at issue, it is apparent to the organization. An area Process Implementation allows you to dig deeper into potential issues from not as visible problems such as expediting costs, long cycle times, long setups, excess inventory, and lost sales.

REACTIVE PROCESS IMPLEMENTATION

If numerical data shows that the end result is out of standard, or you are reacting to something on the floor, then your Process Implementation is reactive. The process for implementing a single process is the same for implementing processes throughout an entire area, but being able to focus exclusively on one process enables you to dig deeper into that process. At least initially, focusing on a single process should only be done for a longer, more technical process such as a machine operator with long cycles, and not for short processes that are quick assembly tasks such as installing one piece to another or screwing in bolts. Being reactive to a smaller single process using Process Implementation can be done, but ensure that the area has a proper foundation before tackling smaller tasks.

Anytime there is a need to implement a new process based on a reaction to something on the shop floor, the cause of this deviation is your aid in developing the new process. Rather than visually seeing that an area needs Process Implementation by not being at the 5S (Sort, Set in Order, Shine, Standardize, and Sustain) goals as shown below, being reactive generally means that an injury has occurred or there has been a quality incident, for example, the defect rate for a process is above target. Solving this particular issue will be your Process Implementation, and the cause of the issue may be that there is not a standard process in place. The steps for implementing a single process from a reaction will be the same as those for doing an entire area. The reactive and area process implementations are only broken out to show that you can do the Process Implementation on a single process or for an entire area.

While working at Toyota Motor Manufacturing Kentucky, I implemented a process in response to the defect rate for bumpers being above target. Each defect was tracked on a sheet, showing both the type and location of the defect, and also the amount of defects per shift. After performing the raw bumper inspection process, and thinking about the process in relation to the defects, I developed a process to both prevent current defects, while also addressing them, i.e., fixing them in process.

For instance, one of the bigger issues was loose plastic from when the bumper came out of injection molding. This loose plastic would develop along the side of the bumper and would get blown onto the bumper surface when it went through the paint booth. Although difficult to detect in a raw bumper inspection, I realized that simply doing a swipe with the thumb down each side of the bumper while wearing a glove would knock any of this loose plastic off, and therefore prevent the defect. I used this same thought process—that is, using the data to determine how the process should be improved, to develop the entire raw bumper inspection process. The new process drastically reduced painted bumper defects, which not only improved quality, but also significantly reduced cost, as it made the entire area of 20 workers more efficient by not having to stay longer for overtime to reproduce defective parts. This example is important not only to illustrate that companies such as Toyota deal with this issue, but also to show that standardized work is not an immediate magic switch for a process that is in obvious need of assistance.

PROACTIVE PROCESS IMPLEMENTATION

If your company has not gone through a Process Implementation like the one discussed in this book, if you have installed new equipment and have had to change the layout of the work area, or if your processes are not documented and visual, then you will need to do a Process Implementation of all the areas affected. Of course, there are other reasons why your company may need to undergo a Process Implementation, but these are the most likely culprits. Even if a 5S and standardized work have been completed, if they are not done following this process and do not meet the following goals, they need to be revisited so that a strong foundation for Lean can be laid.

If you are new to process implementation, start with an area that has identical or near-identical standardized work to get accustomed to the implementation process. Basically, look for an area where the process doesn't require much movement and the standardized work doesn't change, or doesn't change much, for a different product on the same line. Implementing processes and sustaining them take time, so make sure that you choose an area small enough that it can be completed within a couple of months, then move on to your next area. The following are the basic guidelines for choosing an area:

1. **Choose an area where the people are working as a team or as a distinct group.**

    This basically defines the area for you. Although this category alone can make the area too broad and too much for your group, it is a good starting point that can be narrowed by the criteria discussed below. This is often set out in how the company or plant is organized.
2. **Choose an area where the standardized work will be the same or similar for multiple workers in the area.**

    Are workers in the area completing the same tasks throughout the day, and are these tasks cyclical? You don't want to stretch your team too thin in implementing the processes. Remember, this process is focusing on standardized work. If you choose an area that has a large number of workers all performing different standardized work, then you will need a large team, or a lot of time with a small team, to implement new processes for all the processes currently being performed. Thus, even when the area consists of people working as a team, number 1 above, depending on your team, you may need to focus on areas within this area where people are working as a smaller subgroup, where their work is similar or one process is dependent on the preceding process.

The number of employees doesn't always correlate to how quickly you can implement the processes; remember it's about the number of processes and not the number of workers. It may take up to 6 weeks to do an area that only has three workers per shift. In some areas, a 5S Project can be implemented in 6 weeks with more than 10 employees per shift. Again, it just depends on the similarity of work and condition of the area.

3. **Choose an area in size based on how many people are assisting you.**

   Again, you don't want to take on a large area with a lot of workers when you only have a few members on your team that can assist you. If you do, you won't be able to complete all the work necessary to adequately complete the implementation process in an area within a couple of months. A large area with only a few workers can require a lot of work based on how the area is used, so even though an area may be small, it may require more work than you anticipate. Sometimes, a small area can require much more work than a larger area; it just depends on how many processes are being performed in that area. If you initially take on too much with your plan based on the size of your team, revisit the plan and make it more feasible.

4. The smaller the 5S visuals and written standardized work in an area, the larger the focus on that area. Put another way, the bigger the gap between where you want to be, the Goal in Chapter 4, and where you currently are.

   It's all about implementing processes, so the smaller the amount of documented standardized work processes and visual controls, the more work that needs to be done by your team to accomplish these tasks. If a 5S has been completed in the area previously, and if standardized work is documented, this prior effort will assist in analyzing the process, and will save time.

3

Get a Team

After you have identified the area, you will need to assemble a team. Identifying the area scopes the project, and many team members will be chosen or assigned based on their knowledge of the area. The size of your team will dictate the pace; having only a handful of people might limit the project's pace and require your team to move slower and focus on smaller areas, which after completion will move to the next smaller area. Ideally, you will have multiple skilled individuals on your team who can assist offline, as well as a talented hourly workforce that can aid in implementing many of the processes. However, not all companies are able to shift their talent to work on Process Implementation, and a company that does not have prior process experience will likely need more direction in implementation. Thus, in the majority of situations, it is better to tackle one area at a time, and not spread your team too thin on multiple implementation projects.

Identify individuals who do not work on the production floor that can assist you. Employees not working on the production floor can allocate more time to the project than those who are dedicated to hourly production work. Ideally, these members of your team will have production experience, and will also have some special skills that will make the implementation of the processes and visual controls more efficient. In addition to these core team members, single out experts that you will need. Initially, these experts should be involved in safety, quality, and maintenance. Safety and quality are pillars that are being evaluated at each stage of the implementation process, and maintenance generally implements the action items to make a process safer or have better quality by changing or installing equipment. There will likely be additional experts that you will need, but these three areas are integral to Process Implementation and must be involved at the outset.

Getting a team at the outset also establishes how regularly you can meet, and therefore affects the time frame in which the Process Implementation can be completed. Meeting for half an hour each day might not be an issue for you, but everyone else might not be able to sacrifice this time. The more frequently you can meet the better; as we will see later, the slower the process goes once it hits the floor, the more likely that it will lose momentum and employee buy-in. Even if the meeting is only for 15 minutes each day, having a daily meeting keeps everyone engaged and prevents both individual action items and also the entire Process Implementation from having any setbacks.

After forming your team, it then becomes necessary to establish a regular meeting schedule and stick to it. Sticking to it means meeting at your regularly scheduled time regardless of the number of people who are able to attend (the agenda can be changed to go over specific items for those in attendance), and not cancelling within 48 hours of the planned meeting. Hold team members accountable for attendance, require a representative to attend that can adequately fill in if someone is absent, and require notice when someone will be absent. The implementation cannot be completed without the group's active participation, so if someone cannot dedicate the needed time, or is just missing meetings, request someone else from that department to be placed on the team.

The expert members on your team do not need to meet every day, so have a specific day each week set aside for your expert team members, e.g., Monday—Maintenance, Tuesday—Safety, where the expert can join the core group meeting. Begin the meeting with any action items for the expert so that they can leave once their projects are covered, and then move on to your regular meeting action items.

Once you have your team, hold a kickoff meeting with the shop floor workers. In this meeting, explain briefly the goal of the Process Implementation, the Process Implementation Process, and affirm that their involvement is both valued and needed for the process to work. More relationship building may be needed when you get boots on the ground in the area, but at least the shop floor employees will recognize the team members and have an idea of why you are in the area.

In addition to having a kickoff meeting to go over the goals of the Process Implementation, train both the team members and shop floor workers on this implementation process. Giving an overview of the process to the core team, as well as laying out the goals to the team, gets everyone on the team moving in the same direction, for the same goal. The training can be as simple as going over each S in the 5S process, and then going over the Process Implementation outlined in this book. This will get all members up to speed on the general overview, but as with anything, the best way to learn is to go out on the floor and do the work. The shop floor workers should also receive a general overview of the Process Implementation. We will address the training of the shop floor employees on standardized work separately in Chapter 11.

Include all shifts of the hourly employees in the area where you will be working. Although you will be working with all the team members in the area while you are on the floor, also include as many of the workers in the meeting as possible. Rotate which employees will come to the daily meeting, if that is the meeting cycle. Including the employees is not only vital to build the necessary trust needed to change the culture, but it also gives you access to the people who are experts in this type of work and those likely to have firsthand knowledge of how to improve it. Actively involving the "shop floor" employees in the process also enables them to have an understanding of the "why" behind everything that is being implemented. Knowing the "why" and being actively engaged throughout the process will drive employee involvement (a pillar of the Process Implementation House) and will make them highly motivated (a pillar of the Lean House). It is one thing to ask employees their opinions while on the floor, but taking the employee off the line to engage them in meetings truly shows how you value them in this process.

Regardless of the number of people, the skill, and the frequency of meeting, it is important to set a deadline for completion, and hold people accountable for their assigned tasks. Without doing this, the implementation will not be given priority by you or the members of the team. To do this, a project leader must be designated, and a detailed plan needs to be submitted for approval by the team to the Executive Team (or whoever needs to provide signoff). The explanation of how to develop the plan will be discussed later, but a project leader must be established at the outset. A project leader should be someone familiar with standardized work and 5S (ideally, a Lean Manager or its equivalent). If your company does not have a specific Lean position, a strong team leader or supervisor will suffice.

4

Assess Current State and Set the Goal

DETERMINE IF YOU HAVE STABILITY

Just as trying to successfully implement pillars of Lean without standardized work would be ineffective, there are prerequisites that are needed for a process prior to Process Implementation. Even though standardized work is where to begin in a Lean journey, you cannot decide one day to implement standardized work if the process is not ready. Stability must first be established so that any implementation effort can be sustained.

In determining whether the process is stable enough for standardized work, the following questions need to be asked:

1. Is the task conditional and is the task repeatable? If the task is conditional, meaning that there is an "if–then" connection to when something is performed, then the task is conditional and cannot be standardized. The fact that a possible occurrence can change the outcome of how the task is performed by definition means that it is not repeatable. Also, if the task is done infrequently, meaning there is not a set time or frequency to which it is done, then it is a standard to follow, and doesn't necessarily need standardized work.
2. Is the line and equipment reliable and is downtime minimal? If the worker is constantly interrupted and sidetracked on issues relating to mechanical issues, then standardizing would be moot.
3. Are quality issues minimal? The worker should not be spending time correcting defects or struggling with effects of poor product uniformity (Liker and Meier, 2006, p. 125).

The general theme across these three questions is that you cannot standardize a process until the process can be shown to have a consistent output.

5S assessment scorecard

Scoring key

Number of issues	Rating level (1–5)
5 or more	1
3–4	2
2	3
1	4
0	5

Date rated:	By who:	Area:		
5S category	Item	Issues found	Score	Notes
Sort	Safety hazards found (e.g., spills, trip hazards)			
	Unneeded items (supplies, inventory, material) found in area			
	Unneeded equipment and tools found in area			
Set in order	Correct places for items are not obvious (visuals and point test)			
	Minimum and maximum quantities are not obvious			
	Items are not in their correct locations			
	Workstations, rooms, equipment not labeled			
Shine	Cleaning materials are not easily accessible			
	Equipment is not kept clean of dirt, oil, and grease, not tagged with needed repair			
	Work area (walls, floors, other surfaces), not free from dirt, oil, and grease			
	Visual controls are not clean or broken			
	Other cleaning issues exist			
Standardize	Checklists don't exist for cleaning			
	Standards are not known or not visible			
	Quantities and limits are not easily determined (point test)			
	Visual controls are not adequate with all necessary information			
	Items are not retrievable within 30 seconds			
Sustain	Number of audits not performed in week			
	Number of items that were not properly completed (5S check, red tag, etc.)			

FIGURE 4.1

5S assessment scorecard. Use the scorecard to walk through the area and list issues that you see. The scorecard key provides the number of issues for each 5S category and corresponding 5S level.

For this process, just like going on a trip, to get where you need to go (your goal), you need to know where you are (your current state). After answering these questions to determine process stability, use the score-card shown in Figure 4.1 to determine at what level the area is in for the different criteria of each of the 5S items. The same scorecard will be used in setting the goal, as it illustrates what 5S looks like for five different lev-els. Give an accurate assessment for each, as the purpose is not to pencil whip the activity but to honestly assess ("Check"), and "Adjust" anything that needs it. When making your plan, the results of this assessment will enable you to know which items need the most resources allocated to them.

In addition to gaining an understanding of the current 5S state of an area or operation, you should also determine the current state for each pillar of your company. If the company has a safety goal, what is the cur-rent injury rate in the area? For cost, what is the current rate for the area, etc. Documenting the current state for these different matrices will better assist you in setting your goal, but it will also guide you in gauging the success of your Process Implementation.

SET THE GOAL

After determining the current state, you must then define your goal for the area. If the process does not have stability, then the first goal should be to attain stability, and a plan should be implemented to attain this goal. For Process Implementation, the first step is to take each of the 5Ss and define what completion looks like, so that you can both sign off on completing the area and also audit the area for being complete. This defined goal will also assist in setting up your action items, as this goal gives you a clear idea of what is required for the area to get to an adequate 5S level; the action items give you the path to reach your goal, like a map gives directions to guide you to your destination.

Use the same 5S scorecard (Figure 4.1) to determine your goal. If you are implementing 5S into a new area, the initial goal should be something attainable. The initial goal should be to reach Level 3 for each "S," which is the baseline for 5S. Within the timeline, a timetable can be made at which point Level 2 should be reached, but the ultimate goal of the process should be at least to reach Level 3. Use the scorecard to determine the cur-rent level of the area, and use the current state level to determine the goal.

Set attainable goals; make sure that the goals are not too easily attainable, but require some stretching of the workgroup.

The following are the goal statements for each of the 5Ss at Level 3. Depending on your company, these goals may look somewhat different, but these are the basic goals on which to build to fit your process. Also included is a brief statement of the importance of each "S."

Sort—**Goal:** Items in the work area are only those that are needed, at least occasionally, by the workers in that area. **Explanation:** Items that are not used are removed from the work area. Floor space is valuable, and items that aren't being used need to be removed to create floor space for those items that are needed, and for the flow of materials through the work area.*

Set in Order—**Goal:** Items are readily available and placed in defined locations, which are marked with the identity of the item and the required quantity. **Explanation:** Once identified as needed, items have a set location, which identifies what the item is and the required quantity that needs to be maintained. When employees no longer have to look for items, they become more efficient, and focus on improving the process rather than looking for items. This organization also prevents mistakes in the process.

Shine—**Goal:** The appearance of the area remains clean, and items are kept in ready-to-use condition. **Explanation:** Team members are promptly cleaning spills and using downtime to maintain their work area. A daily cleaning standard should be set and the employees should do a quick cleanup and inspection of their area at the beginning and end of their shift. Employees should leave items in good condition at end of shift, and employees should ensure that items are in good condition before starting their shift. A cleaner workplace is safer, has higher quality, and improved morale.

Standardize—**Goal:** Standardized work is documented, and key steps of standardized work are made visual throughout the work area through visual controls. Standards for 5S are also established to maintain the first 3Ss. **Explanation:** The steps for each process are documented, and visual controls are set up for the employees to prevent mistakes and to

---

* Many companies perform a "red tag" process for these items. In the red tag process, you literally place a red tag on the equipment and move it to a specific area where it will stay for a set period until it is used by your area or another area, or is discarded. If there is no red tag area, create one and move any unneeded items to this area.

enable new employees and others to quickly recognize the key steps of the standardized work (No more asking, "what's that?" or "how do you know?" and employees will be able to point to something rather than answer these questions once visual controls are set.) The 5S is standardized and integrated into the final standardized work. The foundation of Lean is to have standardization and visual controls; thus, to set your company on the Lean path, it first needs standardization and visual controls.

Sustain—**Goal:** Each shift is leaving the work area to the 5S goal level for the next shift, and audits are being performed on the work area. Systems are put in place to audit standardized work processes. **Explanation:** Tools are put in place to create accountability. Whether the responsible person of each shift gives a shift update face to face, or whether a logbook is used, each shift starts and ends their shift with everything to standard. Leadership is also auditing the area to ensure 5S and standardized work compliance. Each shift should audit the last shift on specified items. If one shift is left "out of standard" from another shift, the hourly operator doing the audit will visually highlight the out-of-compliance abnormality.

As will be discussed later, teaching and training the workers in the area is the key to sustainment. Without the sustainment piece, the work that is done in the first 4Ss is meaningless, as systems put in to sustain these are critical in maintaining standardization and 5S.

For the 5S, standardization establishes the 5S work standards for maintaining the 5S condition, the goals established for the first 3Ss. Sustain holds people accountable for maintaining this standard. The goal in standardization of establishing documented standardized work through 5S with visual controls is the new aspect that this process incorporates into the 5S. This standardized work is sustained by holding people accountable for maintaining their standardized work through audits and coaching.

The goal for standardized work should be:

1. Fully documented standardized work for all work processes
2. Training program established and implemented for new employees
3. Multi-Function Worker Charts established (discussed in Chapter 11)
4. Audits to coach employees and improve standardized work
5. Countermeasures identified and followed through for all issues

If an area of your company does not match up to the goal stated earlier, then you are likely in need of Process Implementation. If you are consistently in the area, then your scores on the 5S scorecard may be higher, and your overall feelings toward doing a Process Implementation may be low. Sometimes working in an area all the time can blind us to the potential improvements, as we have gained the necessary understanding of how everything works. However, this can cause a "can't see the forest for the trees condition," as we are too close to the situation to be able to recognize needed change. To do a quick assessment after filling in the scorecard, ask these questions:

1. Can someone from a different area of the company watch a process and identify every item without asking questions?
2. Are the steps to how to complete a process being taught to new employees in the same way?
3. Are process steps posted in the area?
4. Are there visuals in the area that point out key steps of the process?
5. Is nonmachine failure downtime less than 30 minutes per shift?
6. Is most of your Process Implementation done by your employees driving improvement?
7. Are quality and safety goals in an area to target?

If your answer is no to any of these questions, then the area is in need of Process Implementation. Answering no to any of these questions identifies a gap in a process or an area.

Point: Remember, it is important to coach the employees on why each step of the implementation is important as you go through the process, so use the explanations of each goal to help the employees understand why the Process Implementation is occurring.

REFERENCE

Liker, Jeffrey and David Meier, *The Toyota Way Fieldbook*, New York: McGraw-Hill, p. 125, 2006.

5

Document and Analyze the Processes

The overall structure of watching the work and analyzing the work simultaneously might seem confusing. This process will be explained in much more detail in this chapter, but the procedure is as follows:

1. Watch the work—watch all work processes to gain an overall understanding of how everything works together and to identify the different processes.
   a. Explain why watching the work processes—to build trust and foster employee involvement.
   b. Ask questions and take notes.
2. Document the steps of each process—the what, how, and why.
   a. Ask questions and take notes.
   b. Is there a standard process: ask what, how, why, who, when.
   c. Do visual controls identify the standardized work, set locations and quantities for all items—focusing on sort, set in order, shine, and waste elimination.
   d. Are there necessary tools or conditions that should be addressed.
   e. Explain why asking questions—to build trust and foster employee involvement.
3. Regroup with team, discuss ideas, get input, make action items against ideas; focus on individual processes but discuss any effect a change may have on a process that pulls from that process.
4. Repeat steps 1–3 for all work processes.

Figure 1.1 shows the process flow for documenting and analyzing the work in Process Implementation. In essence, you are looking for waste with your eyes and from the words of the employees as they describe the process to you. In addition, the Sort, Set in Order, and Shine aspects should be considered.

WATCH THE WORK

As an initial activity to gain an understanding of the area and the processes, you will determine what processes are currently being performed. If the area already has documented standardized work processes, you can use them as your guide to begin documenting the names of processes. However, having standardized work documented does not necessarily mean that processes are implemented, and you should still follow the steps described below.

Before you tackle documenting the steps of the processes, you need to identify the processes. A process involves someone performing a step or a series of steps to achieve a result. If there is an output, then there is a process. Not every single process that is performed at your company will need to have standardized work written for it; the general rule is that a process that needs to be standardized is one that "touches a pillar"—meaning that if it affects safety, cost, quality, or service, then it needs to be standardized.

When defining processes, it becomes more difficult for work areas where one employee could perform multiple processes while working in the same area. So, initially, if you feel overwhelmed, just get the names of the processes, and have a general idea of how they work. As you begin to communicate with the employees, be sure to reinforce to the employees that you are trying to understand their area to better assist them (the "why"). Although you gave an overview at the kickoff meeting, give the employees a rundown of the Process Implementation, and be very clear what will be occurring in their area—that you are there for improvement—and would love to hear any improvement ideas that they have to make their process better.

Go to the Gemba, where the work is being performed, and talk to the workers performing the process. In observing the current work that is being performed, this might be your first occasion to visit the process and talk to the employees performing the process. Even if you are familiar with the process and employees, the way in which you build trust is vital in ensuring success in implementation, so think back to the communication principles from the Culture section in Chapter 1. As you can see from the Process Implementation House, the foundation is employee coaching; within this lies constantly engaging the employees into "why" you are in their area, and the reasoning behind what you are doing. Employees must understand the importance and reasoning—"the why"—in order to make improvements; not only is explaining why decisions are made vital (Mayfield and Mayfield, 2006), but you can't improve a process you don't understand (Adler, 1992).

The employees will have a name for the process and be on the same page for how the process should be completed. For now, just document what name they give you for the process. Ask questions about what you see (e.g., what do you do if…, where do you get your supplies from, does this process ever require you to do something extra). Some parts of the process, or whole processes, may be performed so rarely in a day that you might not see them if your time in the area is limited to a certain time of day. You will perform this practice again when you focus on each specific process, but having an initial big picture to facilitate the 5S project and gaining an understanding of the area is necessary at the outset.

To analyze the process, you must be able to determine what the processes are. Many manufacturing companies have employees that perform the same process for 2-hour increments, repeating the same process over and over for that period. For processes such as this, standing and watching the process for only a short time will make it evident what process is being performed. However, not all processes are so easily recognized. Although this initial activity allows you to gain an understanding of all the processes, you will undoubtedly begin to see things about individual processes and ask questions about how the process is being completed; being able to gain some specific insight into a specific process will only be a benefit, just be sure to take notes of what you see and what the employees tell you. You will see some carryover questions from this section throughout the process. The questions from when you are watching the work are for guidance. Being in the area and watching the work will drive some of the questions; just don't get too deep in the weeds on all of the processes as you are really just getting the basics at this point.

Create a process flow like that shown in Figure 5.1. Having a sheet that shows how everything works in sequence provides a nice visual and also provides a single list of all of the processes in the area, which can be used as a checklist for completing Process Implementation on each. The processes that need to be addressed are those involving set up and shutdown (Pre-Production set up, Pre-Production Per menu, and Post-Production All Menus in example below), in process procedures (e.g., line worker process during shift "Production" processes in example), and a cleaning schedule (listed in example in Post-Production). For processes that are not part

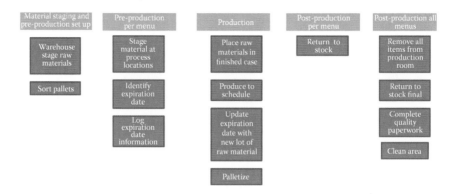

FIGURE 5.1

A Sample Process flow that shows the relationship of the processes in the area. Just creating the process flow may show gaps to issues that need to be addressed ("What happens when..."), and also provides one document that shows the interrelation of the processes to one another. The process flow also identifies what the processes are and what needs to be trained and audited.

of production, a process flow can still be created; it may just not follow the same structure as that just outlined. As noted previously, if a process touches a pillar, then it should be standardized and should have documented standardized work.

Once you have identified the processes, you are ready to begin documenting and analyzing.

DOCUMENT THE STEPS OF THE PROCESS AND ASK QUESTIONS

Documenting the steps of the process and analyzing the process should be done simultaneously, as asking questions while you are on the floor watching the process is when questions will arise, when you have the experts available to answer your questions, and you are able to visually identify waste. Always ask questions, but don't question. No one wants to hear you tell them they are doing something wrong or question how they are doing something, especially if they are doing the process the way it was explained to them. Think of this activity as information gathering, not a cross-examination—more *Columbo* and less Tom Cruise in *A Few Good Men*. As you are on the floor, continually ask what, how, why (which will make up the standardized work sheet), but also ask who, when, or any other question that will help you to fully understand the process. Ask these questions for both processes that you see, and anything that you don't see: you don't do this; what is that tool used for; do you ever have to, etc. Before implementation is the best time to get all of your questions answered, but do so in a way that you establish or maintain trust—that is, again, ask questions but don't question.

Not only should you focus on coaching employees so that they understand and trust the process, but engagement is also required to develop continuous involvement. Thus, you need to ensure that a culture of trust is established so that employees feel free to express their opinion. According to research by Amy Edmondson, professor at Harvard Business School, and James Detert, Cornell University professor at Johnson Graduate School of Management, two beliefs are essential preconditions for the free expression of upward voice:

1. The belief that one is not putting oneself at significant risk of personal harm (e.g., embarrassment)
2. The belief that one is not wasting their time in speaking up (Gilbert, 2006)

Even in work environments conducive to employees speaking up, people can still be afraid to speak up to their boss (Gilbert, 2006). Listening to employees ideas and walking through each idea becomes critical in not only building a culture of trust, but it also encourages employees to speak up, as they see that their ideas are actually being considered. Every manager needs to work at being open and accessible and taking action on ideas or reporting back on why action can't or won't be taken (Gilbert, 2006). Make time for listening to the employees; each will be different in how much they share, as some will want to talk to you nonstop whereas others will likely only give information when prodded. Giving time to each employee will take some additional time, but the investment is well worth it.

Observe how everything works together in this step. Watch the work, document the steps, and start taking notes. You are still in the "Planning" phase of the Process Implementation; you will begin to move items around once you enter the "Do" phase. In conjunction with documenting the "what, how, and why" steps of the process, document every thought that you have related to what you observe while you have it or you will forget it, and make sure that you fully explore all ideas. The notes that you make are what will drive your action items as you begin implementing. As you are watching the work and writing down the steps, look for all forms of waste; focusing on one process at a time is beneficial in allowing you to focus on all of the details of a process to identify waste and not become distracted. Watch all shifts perform the process and communicate with all shifts. Including all shifts on the shop floor not only builds a trust culture and elicits employee involvement, it also enables you understand how each person performs a process.

Begin to focus specifically on documenting the steps of the process, as they are currently performed, for all the processes that you defined in the previous step. Even if the steps of the process are already documented, write down the steps as the shop floor employees describe and as you see them, meaning that if they do something but don't address the step, clarify what they are doing and add it to the document. Analyze the process while watching it, and once you are reflecting, pull up the current standardized work steps if they exist and see if they match the steps the employee walked you through. If something is different, reflect on why it is different and determine which step is better for Process Implementation (i.e., using the pillars of the Process Implementation House). If workers are straying from the current standard, determine the reason and deal with it accordingly— e.g., if they did not know standardized work or thought their way was better, then coach them. Employees cannot unilaterally change a process, but must run their ideas past their supervisor. Standardized work is the best way that your company knows how to perform a job today, and everyone must follow it while it is the standard. Often, the standardized work process has been changed, but the standardized work sheet has not been changed, which is a gap that must be properly addressed; the document must match the reality of what is being done.

There is no magical sheet for documenting these steps, so long as you are writing:

1. What the step in the process is
2. How the step is performed
3. Why the step is being performed*

See Figure 5.2 for an example of a standardized work sheet. The most important element of this exercise is that you write the standardized work

---

* Eventually, this information should be formally documented on a standardized work sheet that is standard throughout the organization. So if you currently aren't using one, adding a standardized work sheet is essential to completing the Process Implementation.

Standardized work sheet

Process name:			Safety		Page of
			Quality		Last revised:
Approved by:			Productivity		
			Cost		
Step #	Work content description (WHAT)	S,Q,P,C	Key points (HOW)		Reason why
1	Retrieve empty batch	QPC	Retrieve empty batch from its lane in the dry room hallway. Carts holding empty batches will be staged with the handle facing the wall		To begin working on batch
2	Complete batch	QPC	See standard work sheet- "Complete Batch"		
3	Stage batch	QPC	Place cart carrying completed batch in its lane in the dry room hallway, with the handle facing the aisle		So cook can identify that batch is completed and can take cart into batch room

Pictures reference step number

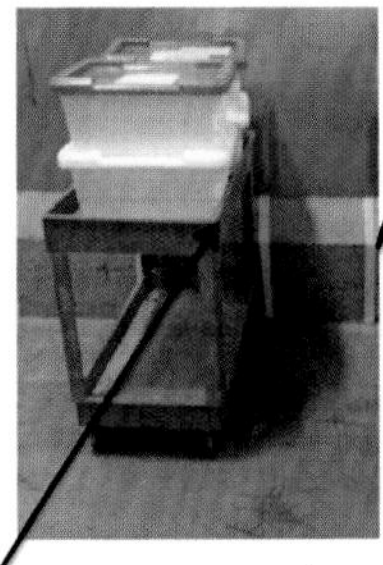

Pictures reference step number

Handle facing wall   Yellow tape marks off lane

Handle facing aisle

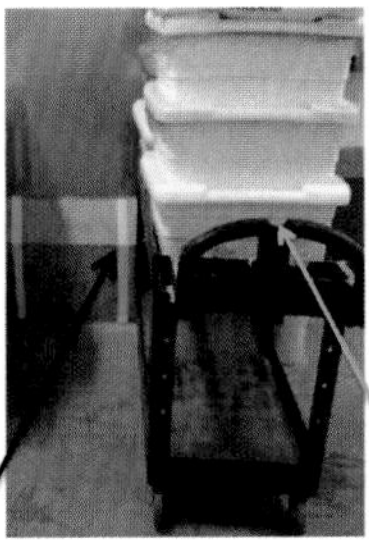

FIGURE 5.2
Standardized work sheet.

with the person doing the work while they are doing the process. This way, you can observe the process while they walk you through it, and you can ask any questions you have about why they are doing something they are doing, or how they know how to do it, while watching the process. Having the employees talk you through the process while they are doing it also gets them thinking about the process, as they must answer the what, how, and why; it should also trigger questions for you about their answers or anything that you see. The steps of the process are usually second nature to the employee, so they sometimes won't think to offer up what they consider the mundane part of their job. Therefore, if they are doing something without explaining it, clarify whether it is a step in the process. It is also at this time that employees will usually have improvement ideas about their process, or open up about other ideas that they have; be attentive and always circle back with them as to the outcome of their idea.

Ask yourself "why" throughout the steps of the process, and don't just ask it once; ask it multiple times to get to the reason the employee is performing an act. Taiichi Ohno, the father of the Toyota Production System, gives the following example of asking why five times:

> Why did the machine stop? There was an overload and the fuse blew.
> Why was there an overload? The bearing was not sufficiently lubricated.
> Why was it not lubricated sufficiently? The lubrication pump was not pumping sufficiently.
> Why was it not pumping sufficiently? The shaft of the pump was worn and rattling.
> Why was the shaft worn out? There was no strainer attached and metal scrap got in.

By asking "why" five times, you should notice a couple things. First, it shows the natural progression of continuing to answer the why for each previous answer given. If you stopped after the first why, you would likely band aid the situation and replace the fuse, without knowing the root cause of why the fuse blew. If you stopped after the second "why," then you would add lubrication to the bearing. As you continue through each "why," you could implement a countermeasure that might fix the root cause short term; but without continuing to ask why, you will not determine the root issue that needs to be addressed.

Although the process is often referred to as the "5 Whys," it may take more or less than five to reach the root cause. In the preceding example, it appears that additional questions could be asked regarding why the strainer was not attached, which may lead to an additional question. The point of the exercise is to gain an understanding of what happened, so ask why as many times as you need in order to gain a clear picture of what happened and the reason for it. An example for the types of "why" questions in Process Implementation could be

> Why did you have to walk over there to get that part? Because the warehouse stages it there for us.
> Why do they stage it over there? Because there is no room in my area.

Why is there no room for it? Because they leave enough for us to complete the day.

Why do they leave so much? So they don't have to keep bringing more.

In the preceding example, you are gaining insight into a potential issue of a worker having to constantly go and retrieve an item because of its staged location, far away from where the work is occurring. After gaining an understanding of the problem through the "why" questions, you will then start to ask the what, who, how, when, and where questions. The rest of the documenting of the work will show you how much is used in each cycle, and thus how much of the product is needed per day. In your mind, you should be thinking about a better way to stage this item so that it is closer to the employee, using the additional information from documenting the steps (explained below), and asking questions to the warehouse regarding their standards for bringing materials (covered in Chapter 6). As you can see, sometimes the issue involves coordinating with multiple departments, which explains why you will need representatives of different departments at your meetings.

When working through the Documentation phase, the focus is on the first 3Ss—Sort, Set in Order, and Shine—while also continuing to keep the principles of culture and waste elimination at the forefront. The following are the relevant items in which to focus on for these 3Ss:

Sort—sort needed and unneeded items
- Ask the employee about tools that are not being used.
- Observe the tools that are currently being used, if the employees have enough, and if there is possibly a more efficient tool.
- Consider whether changing or adding a tool would make the process more efficient and safer, or improve quality.

Set in Order—a place for everything, everything in its place
- Find out each item's specific use and frequency of use.
- Determine ease of access for necessary tools.
- Focus on visual controls: Are defined locations labeled? Can employees explain how they do the process by pointing at a visual control, if feasible?

Shine—keeping the workplace clean and safe
- The necessary cleaning supplies are readily available.
- A cleaning schedule is established.
- Are employees work areas set up for each shift?

As you listen to the employee talk you through the process (the what, how, and why), look for keywords that identify that there really is no set standard or visual management for the process. Usually, these are identified with the following phrases:

"It usually"
"Most of the time"
"I do it this way, but they do it this way"
"We just know"*

Employees are basically telling you that the way in which they do a process changes and/or it changes when performed by someone else. Phrases like this should immediately send up a red flag that a process needs to be implemented that everyone will follow every time.

Start with a simple process that only has one or a few steps and doesn't involve much movement. For the what, just write down the simple step that the employee performs; the how is where you describe the detailed steps of how to complete the "what." For example, if the employee was doing a process that required pushing a certain number of blue clips into certain holes, you would want to limit the "what" to something like "apply blue clips" (or however the clips are labeled when the employee gets them), whereas the how would describe how many clips to retrieve and from where, how to apply them, and into which holes they should go. Check with the employee on why they are doing this, but in this example, common sense will likely tell you that it is to secure something on the end product. Applying these clips may not be the only part of the process, but if there are more steps, then they should be documented in the same fashion. Basically, be sure to keep your what part simple, with the "how" being the meat of the documented standardized work.

As you are performing this exercise, write down the steps exactly as they tell you and watch them go through the steps. In addition to the what, how, and why, ask questions about anything that looks out of the ordinary as you watch the process, focusing on waste elimination. Make notes next

---

* "We just know" seems to be the most common of these phrases, and groups that have been doing the same thing for a long period will tell you that Process Implementation is worthless because everyone knows how to do it. Again, if an outsider doesn't know how to do it, then the Process needs to be implemented.

to the step in the process that is inefficient, focusing on the following types of waste:

1. **Overproduction**—Are employees "pushing" the product to the next process, meaning that they are building up too much work in process?
2. **Inventory**—Not very easily seen when observing a process, but the workspace may be too crammed with work in process inventory that can be removed and be transported to the floor throughout the day.
3. **Transportation**—Usually caused by excessive inventory or poor layout; moving items for production should be done quickly (point A to point B) and efficiently (single handling).
4. **Motion**—Anything that looks inefficient. Documenting everything that you see while watching the process makes finding wasted motion more apparent. Use common sense in watching the body movements of the employees as they perform the process; employees often have ways to improve the movements of their process.
5. **Unnecessary Processing**—Again, documenting everything that you see while watching the process will make waste become apparent. The documented work steps will show unnecessary processing, as will engaging in the employees and asking them about what is required to complete the process and what issues they have.
6. **Defects/Quality**—Someone internally should be tracking all defects and quality issues in the area, so that metric will be your guide.
7. **Waiting**—Anytime you see individuals waiting, engage the employees as to what they are waiting on. Sometimes employees do not want anyone to notice that they have some down time, so approach with caution on letting the employees know that you are on to them initially.
8. **Underutilizing People***—The foundation of the Process Implementation House is employee coaching, and a central pillar of the Lean House is employee involvement. Improvement ideas at your meeting should not only come from the off line team members, but also from the shop floor members.

If you documented the steps of the process accurately, and analyzed the employees' responses to the what, how, and why, you should gain an

---

* There were initially only seven types of waste; Jeffrey Liker introduced the eight types of waste in *Toyota Way*, 2004.

understanding of what items are needed and whether they have a defined location. While working at the food plant, I was documenting the steps of the employee retrieving dry ingredients process. My notes on the process looked something like this:

What: Retrieve metal cart

How: They are usually in that room, so we go and get one and bring it back to this area to get our dry ingredients.

Notes: Carts not in set location, carts have no defined location; we currently have four carts, is that enough, where could we locate an area for carts near where the dry ingredients are loaded so employees don't have to go to get them?

Why: To load dry ingredients that we take into our room.

Writing down this process, asking questions, and analyzing what was needed showed a couple things. First, the metal carts did not have a defined location. After documenting the "how" of retrieving a metal cart as "retrieving one from the general area," I made a notation on the standardized work sheet that metal carts needed a defined location. This also showed that there was wasted movement. The metal carts were not being stored in the employees' work area, as they had to "go and get one" and bring it to where it was needed.

Using reflection, the process discussed in the following chapter, we took the idea of shopping cart lanes at a store to make a defined lane next to the work area, where employees could gather a cart and take it only a few feet to the area where it was needed. This was a very small piece of the overall Process Implementation in the area, but it shows how each small task should be analyzed, as this change made this individual process much more efficient. Thus, the "how" of retrieving a metal cart became, take one from the metal cart lane, and employees could point to the lane where they were retrieved, as shown in Figure 5.3. Thus, not only did the metal carts have a specified location ("Set in Order"), but visual controls were established to easily identify this space. When visuals are in place to help employees perform the "how," they can point to the visual when asked how they know to perform the step, rather than giving a detailed description. Each "how" should be this definite.

As another example of having a set place for everything, Figure 5.4 shows a designated location for drum dollies.

FIGURE 5.3
Lane established for storage of metal carts when not in use.

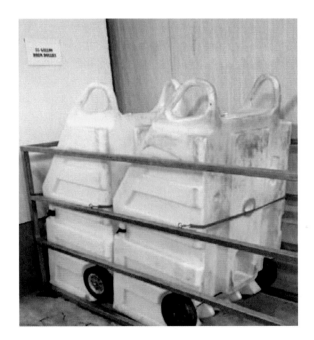

FIGURE 5.4
Drum dollies lined up in a defined location.

The majority of processes involve the employee lining up different items, and/or setting the finished product in a defined location. In these instances, there should be a visual indicator assisting the employees to line up the item they are using with the equipment, and would look something like:

What—Line up A with B
How—Line up yellow mark on A with yellow mark on B

In this example, if asked how they line up A with B, rather than giving an explanation of the process they use, they can simply point to the yellow marks and give a very brief description.

When I worked at Toyota, one part of a process in my group involved attaching a bar code sticker on the inside of the raw bumper, so that the paint booth scanner could read the bar code and apply the correct paint. When new employees asked where the label should be placed, we could point to a molded square inside the bumper, which was applied to the bumper during the injection molding process, that identified the placement of the sticker, rather than having to describe to new employees how far up and how far in it should be placed. When you can point to the "how," it eliminates confusion.

In another example, while documenting the steps of a process at a food plant, the employees described a process as follows:

1. What—We line up the lift in front of the kettle that we are getting ready to use ("Line up Lift").
2. How—We push the lift until we get close to the kettle and then get in front of the lift and make sure it is lined up in front of the kettle.
3. Why—To line up the lift in front of the proper kettle.

With the language used, and some questions (e.g., how do you know when it's lined up in front of the kettle), we took the issue back to our meeting and discussed how the process should look. We came up with the idea that to create a standard, we could apply tape to the rail that the lift rolled on, marking off where the wheels should be when the lift was exactly in front of the kettle. To help in the implementation, we assigned a shop floor employee in the area to determine the correct position and to apply the tape. After the implementation, the process looked something like this:

1. What—Line up lift in front of kettle
2. How—Push lift (push lift was more in depth due to ergonomics of the safety issues in actual standardized work sheet) until wheel is between taped lines
3. Why—To ensure that lift is lined up in front of kettle

Figure 5.5 shows the wheels of the lift lined up between the taped lines. As the employees would push the lift from the side, and therefore be next to the rail, these lines not only made the process of lining up the lift standardized with visual controls, they also made it efficient by having the visual indicator easily identified.

Something else that we did with the rail taping idea was to assist the shop floor employees in developing this idea. Although we quickly knew that this inexpensive idea would easily enable the employees to execute the "how," employees often gave us crazy looks when we first began discussing marking the wheels for dumping into the kettle. To create ownership of the process, we went over the steps of the process with the employees and threw out our general idea for making the process better, taping the location of the wheels. Having them think through the process and come up with the exact idea will create ownership and assist in accountability,

FIGURE 5.5
Two pieces of tape identify where the wheel should be located when lift is centered in front of equipment.

as they are much more likely to enforce the process if they think it is the best way to perform the process. It also begins to motivate employees, an essential element in operating Lean, as they are empowered to think of ways to implement new processes because they see that their ideas are put into practice and make their job more efficient.

If you have worked at a company a long time and know where to locate items and the identity of the items and equipment, try to put yourself in the shoes of someone who has never set foot in your facility. It is good to bring in an outsider who doesn't have experience in the area to see what needs to be identified, because if they have to ask the identity of something, then it needs to be identified. In examples such as those cited earlier, it was easier for me to identify when visual controls were needed because I was new to the area and, through Process Implementation, was constantly asking "how" employees knew how they needed to perform a task.

Another similar example from a food production facility involved how carts that carried mixed dry ingredients were being staged. During the "Watch the Work" phase, we noticed that the plastic carts carrying dry ingredients were placed outside the room where they were being made that day, which changed daily. Remembering this observation when documenting the steps, we went into the "Document the Steps" phase knowing that the carts did not have a set location. However, by following the guidelines discussed earlier, we documented the steps exactly as the employee told me for this process.

When documenting the steps of the process for completing a set of dry ingredients (e.g., mixing starch, flour, salt, sugar), the employee said that the first step involved retrieving a cart of empty tubs from outside of their room. When asked how they knew which ones needed to be completed and which ones were already done, they responded that they just knew because they had done them, basically saying that they kept a running log in their head of which ones they had done, and which ones still needed to be completed. When I inquired as to how someone that had not done them could know, they said they could lift up the lid of the tote and see whether it had any contents.

Thus, the "how" for retrieving a set of empty totes for dry ingredients was "go outside room and retrieve cart." My note on this step was that the employees either knew which ones were empty or they looked under the lid to see if the tub was empty. As mentioned previously, "we just know" is a red flag statement of a process not having a standard, as was the case in this process. In this particular example, the process also had safety and efficiency issues, as employees had to maneuver through the carts to get in and out of the room, and it took time to determine which totes needed to be completed, both for the employee filling the tub and for the employees taking the tub for use. Basically, analyzing this process using the above framework led to many questions and many ideas for improvement.

The planning aspect of this particular Process Implementation lasted for days. In determining where to stage all of the plastic carts—and therefore where to move any items that were currently in the hallway—we had to first determine the amount of plastic carts that are needed in a day. Once this was established, we had to research what raw material items were staged in the hallway, and their purpose for each being staged there. After finding a potential space for the plastic carts, we went about measuring the entire area to ensure that our plan would work, and ran our idea past the employees. Next, we diagrammed the area with the new layout, showed our proposed plan to the team, and also showed it to the employees to get input. Although we were able to find a set location, and identify those locations with marked lanes, there was still no identification showing which cart of tubs needed to be completed by the dry room employee, and which cart of tubs was completed and ready to be taken. To create a visual cue for empty and completed carts, we simply determined that carts that are completed should be staged with the handle facing the aisle, and carts that need to be completed should be staged with the handle facing the wall (Figure 5.6). Not only did this create the visual cue, but it also created a safe process, as the completed carts were heavier and therefore needed to be pulled by their handle, whereas the empty carts were much lighter and could be pulled from the wall with ease. This example shows how using both 5S and waste elimination through standardization (i.e., Process Implementation) can be very effective.

After implementing the process, the steps for completing a cart of totes and retrieving a completed cart of totes both involved taking them from their lane in the hallway while following the identifiers, handle facing in or out (which was also referenced on the wall above the area; remember to identify key steps of the standardized work through visuals so that you can point to something that explains the "how").

The examples discussed in this chapter show the entire Plan–Do–Check–Adjust (PDCA) process, including the implementation of the process on the shop floor. Although the overall Process Implementation process is a PDCA cycle, each step within the process also has its own PDCA cycle. The only action that needs to take place in this phase is watching the work,

FIGURE 5.6

Both carts in the above picture are complete, as evidenced by the handle of each facing out. Refer to Figure 5.2 to see how this process would look in written standard work.

documenting the steps, and analyzing. The examples used in this chapter show how the ideas were carried through, and will make more sense after reading the following chapters.

REFERENCES

Adler, Paul S., The "Learning Bureaucracy": New United Motor Manufacturing, Inc., School of Business Administration, University of Southern California, p. 59 (1992).

Gilbert, Sarah Jane, "Do I Dare Say Something?", Working Knowledge, Harvard Business School, March 20, 2006.

Mayfield, Jacqueline and Milton Mayfield, Leader Communication Strategies; Critical Paths to Improving Employee Commitment, American Business Review, p. 90 (June 2002).

6

Meet with Your Group
and Assign Action Items

After you have documented the process with the employees, asked questions, received employee feedback, and analyzed the process for other concerns, reflect on the process both on your own and also with your group. Taking some time away from the process will help you see the forest for the trees and refresh your thought processes. In addition to any brainstorming activity that you do individually to get ideas, discuss the notes and documented processes that were written while you were on the shop floor with your team.

MEETINGS

The cadence for the team meeting should have been established in Chapter 3. Once you begin the Process Implementation, these meetings should be taking place and have consistent participation. In these meetings, take what you have learned from being at the Gemba and your notes back to the team meeting. Not only should you be working on obtaining buy-in on the floor when discussing potential ideas, but the team meeting is the forum where all shifts are represented and ideas can be discussed. You've already established your plan to attain 5S and standardized work goals, but once you begin the actual process, you will use your meetings to develop action items, which will drive different tasks to be completed by specific people (i.e., maintenance issues should be addressed by maintenance, safety issues by safety). The shop floor employees will hopefully be regularly attending the meeting and raising different ideas as well (remember to walk through all ideas that an employee brings up, talking through their feasibility). Ideas that you have while watching the work, documenting the steps, and talking to the employees should be driving a lot of the action items. Having the shop floor employees more directly involved in meetings expedites the process, as although you are working with the shop floor employees by documenting the steps and analyzing what is occurring, employee comments can drive discussion for action items in the meeting, which will only help increase the effectiveness and efficiency of the Process Implementation.

Reflect with the group about the current standardized work by going through the current standardized work, if any, the documented process that you recorded in Chapter 5, and any notes that you took on the process. From your notes, start assigning action items to members of the group based on the items that need to be changed for one of the reasons that you looked at above (e.g., look at making part of process safer, part of process not standard, standardized work not specific, needs visual controls). The action items don't all need to be specific; it could be assigning someone to solve an issue, such as finding a better way to perform a process because employees are lifting over their head too much. Eventually, solving this issue will involve specific action items, but assigning the project can be general initially. If you have come up with specific things that you would like changed, go over it with the group to see if it is feasible, get an idea of the cost and how it would potentially affect the area, then assign specific action items to get some assistance in the implementation.

Be mindful that you may get some pushback for your suggestions about making the process more efficient, as noted earlier. However, continue to reinforce the benefits of implementing the new process and the overall goal for the Process Implementation of the area. After the employees begin to see the positive changes to their area, it will be easier to get their buy-in for implementing new processes.

Before rolling out a new process or a change to an existing process, it is best to think through all the variables and have the best process at the outset. Although you could "plan" less and "do" quickly, workers do not want to deal with an inefficient process that will need to be continually adjusted. If you put something out on the floor at the outset that is a much improved process, workers can begin to familiarize themselves with it, and not have to deal with the process potentially needing an overhaul soon after initial implementation, and then having them learn a different process.

In the examples given in the previous chapter, although the initial idea came about while at the Gemba, the ideas were discussed at our daily meeting, someone was assigned an action item for completion and given a date when completion should be done. Even if you have an idea and can quickly implement it, not only does the group need to be aware of the action so that they know what is taking place in the work area, but someone in the group may have an idea to improve upon yours.

As you talk through some ideas, reflect on instances outside of the shop floor on which you are focusing. I was recently at the Department of Motor Vehicles (DMV) to transfer the title of a car. To complete the transfer, I had to bring the notarized title and a copy of current insurance on the vehicle. After filling out the remaining information needed on the title to transfer, I paid the fee and began to leave. As I hit the door, the woman helping me yelled for me to come back. She realized that I had accidentally given her the expired insurance. We easily solved the issue by calling the insurance company and having them fax over the current policy. However, if I had already left before she noticed this error and the insurance company would not release this information without my approval, then I assume that the only way to notify me was to send a letter to my house requesting that I come back in with current insurance to complete the transfer, which would have been a real headache.

I had this example in my head as we went through the "sort" exercise, as it showed how easy it would be to mistakenly use an item in the work area that should not be used. Once the old insurance was expired, I should have thrown it away. This way, I would have prevented myself from making this mistake, as the only insurance I could have taken was the current insurance (as in sort, throw out things that are not needed). For the DMV, they could have a check sheet for each task that they perform. In this instance, it could have been a checklist of items to check both on the title and the insurance for a title transfer, i.e., what information needs to be filled in, match the vehicle identification number (VIN) on the title with the insurance, and check the insurance card for current date. The purpose here is just to think about real-life examples and what you could have done to prevent them, or actions you took in advance that prevented the issue from occurring. The takeaways from my example with the insurance card didn't immediately assist me in implementing a certain process, but these real-life examples were in the back of my mind as I went through the implementation process.

Reflection is a good way to keep a process mindset; remember, the point of reflection is just to be observant of the processes around you and to be mindful of how things in our everyday life, or even watching a different process, can assist in something that you might be doing in a Process Implementation.

Section II

Do–Check–Adjust

By the sheer content of the Plan section compared to the Do–Check–Adjust–Section, you can now visually see most of your efforts in planning. Adequate planning should be done before entering the "Do" phase. This is not to say that there is still no work to do, but once you reach this phase, you should implement rapidly.

7

Implement a Pilot

When implementing something new, initially just perform a trial of the process, often referred to as a pilot. A pilot is a small scale version of a larger project that you are working to implement. Among other things, doing a pilot allows you to catch potential problems in the process prior to a full scale rollout. Although when implementing a pilot for a process, you are technically in the "Do" phase, the pilot allows you to "check" whether the pilot is successful and "adjust" if necessary, which will trigger the Plan–Do–Check–Adjust (PDCA) cycle once more.

The steps for performing a successful pilot are as follows:

1. Engage employees about how the pilot should be set up—Basically, the "plan" phase from above, as the goal of that phase is to take ideas from the floor and your groups ideas, and mesh them into a new improved process.
2. Engage employees about how the pilot should be changed—Once you roll out the pilot to the process or area, have the workers that will be performing the process perform the pilot process and ask them for suggestions (the same way that you did earlier in the initial plan phase).
3. Implement the idea on a small scale
   a. One tool or visual control if these will be rolled out across the area
   b. One line if the pilot line will be rolled out across the plant and/or different plants
4. Adjust the pilot if necessary and repeat the PDCA cycle for the pilot.
5. If the pilot line is successful, roll out on a full scale (see next chapter).

Pilots are beneficial for the following reasons:

1. It allows you to confirm process is ready for full-scale implementation—While analyzing the process in the pilot phase, you are able to determine if any adjustments need to be made to the process. Watching the process allows you to "start over" in essence, because you are watching the work and analyzing the work in this phase much like you have already done.
2. It prepares the employees for the process when it is implemented on a full scale—Allowing the workers who will be performing the process to work side by side with you on any tinkering that needs to be performed, as well as begin to learn/perform the process under nonproduction stress will enable success when implemented. As mentioned previously, this is why it is important to put a good process out initially; employees don't want to waste time on a process that does not meet their needs and have to be continually learning new processes.
3. It engages the employees, allows for coaching, and creates an opportunity for buy-in—Process Implementation begins with Employee Coaching, which creates employee engagement. Coaching the employees creates confidence and builds trust.
4. It enables buy-in from the employees—Walking through the pilot with employees who were skeptical of the Implementation benefits of the new and/or improved process. You can only preach the benefits for so long before you show the benefits.

The purpose of a pilot is to implement a trial to one line to ensure its success, prior to rolling it out on a larger scale. Not only does this ensure that the best process is implemented, but it saves cost in that you will be able to recognize issues from only implementing once and on a very small scale, rather than a full-scale implementation. In Process Implementation, you may at times be implementing a pilot for only part of a process, or you may be implementing a pilot that will eventually have an effect across the entire organization. As an example of a pilot for part of a process, during a Process Implementation, we recognized that employees needed a paddle in order to scrape food contents from a stainless steel cart. Although we did the proper planning to determine some of the parameters (it could only be a certain length because of space constraints in the work area, it needed to be metal detectable, the item must be food grade, it should be cleanable, etc.), we purchased two potential paddles for a pilot, which were discussed at one of our group meetings with the team, including quality and shop floor employees. Since the paddle needed to be sanitized prior to its use for the day, we spent a couple weeks allowing different employees to use the paddle for the entire day on their line. With the employees' feedback, we selected a paddle that was purchased for all workstations (10 in all).

The benefits that we realized from the pilot were small in the overall Process Implementation, but multiply this one instance by the amount of implementations in the area, and the overall benefit of the Process Implementation was immense. In this example, we obviously saved money and time by not only doing some initial planning, but by only purchasing a total of two paddles for the employees to try. Although they preferred the paddle that we assumed they would, performing the pilot showed them our commitment to finding the right paddle for them and helped gain their buy-in before rolling out the decision. Once the paddles arrived for all workstations, the employees already knew how to properly use them.

We received some initial pushback to implementing the use of these paddles, as employees felt their way of extracting food from these carts was effective. As with taking employee suggestions (discussed earlier), we addressed each suggestion that they brought up, but always firmly defended the process if it followed the pillars of the Process Implementation House. If you laid the groundwork with the employees in the Plan phase (as noted earlier), then they should all have a general idea of the new process and why the process is being changed, and therefore pushback should be limited in this phase.

After you think you have the process finalized, make temporary visuals* for the process and see if it passes the "Point" test. For a completed process with visual controls, you should no longer be unable to identify items or where to find them. When you ask employees how they know what something is, or ask them how they know how to do a key element of the process, they should be able to point to a visual that either identifies the item or describes how they are to perform the process (remember the example of lining up A and B with the yellow line discussed in Chapter 5). Another good way to test whether the pilot process is ready to be rolled out is to bring in an employee who is not familiar with the work area and have them attempt the process by reading through the standardized work

---

* Don't spend too much time making identifiers for the process; just make the temporary visuals as simple as they can be. It is when the process is finalized and implemented that you should make your permanent visuals.

that you have written and use the temporary visual controls that are in place. This could be another forest-for-the-trees exercise; your team and the employees may be too close to the process to identify needed improvements, but if the goal is to make the process so that anyone can perform the process without asking questions about the "what" or "how," then it's a good exercise to have an outsider try to perform it.

Part of this Pilot process involves identifying the minimum and maximum amount of supplies needed, where they should be staged, and who will stock them and check the contents daily. The idea of having the minimum and maximum is that the employee should not have to hunt for an item during production, but rather, each shift should stock the maximum quantity at the beginning and end of their shift, and during any downtime. As mentioned before, the minimum and maximum do not need to be addressed in the standardized work, but are standards that must be trained.

The minimum level should be enough to allow time to continue to operate the process while the item is being retrieved. The maximum will vary depending on the type of item, but be sure to allocate enough so employees are not continually getting more supplies, while also considering the available space you have to stage items. The area holding the item should be marked with the numerical maximum and minimum quantity. Another way to mark the minimum and maximum is with lines or other types of markings showing the minimum level and maximum level. For example, if items are stored in a tray, you could mark a green line showing the maximum height the item should reach in the tray and a red line near the bottom showing the lowest level the item should reach. When the item reaches the red line, it should be refilled to the green line. Either way, you should be able to look at where items are stored and know whether they have too many, too little, or enough. If your process passes these tests, then it is time to move to Implementation.

8

Implement the Final Process, Share Success, and Begin the Next Process

Following your additional planning, implement any changes that result, and again go through the process of watching employees perform the process, gather notes, and get employee feedback. For all of the examples that were given earlier, we implemented a pilot and went through the Plan–Do–Check–Adjust (PDCA) process to determine whether to roll out the initial pilot or to make changes to the pilot and repeat the process. If the pilot is successful, you should implement the full-scale process.

VISUAL CONTROLS

Once the pilot is ready for a full-scale rollout, finalize your visual controls as well, as these should have been piloted concurrently with the process. For items that are permanent, such as a piece of machinery or a tool, make a permanent label or identification. For locations that can hold different items, or storage that holds different material, you will need to make flexible identifiers that can be changed instantly when the item or material changes. As discussed previously, if you are uncertain about the final staging for an item because it may be affected by another process, then leave your temporary labeling up until you complete the processes that may have an effect on this process.

The purpose of visual controls is to easily identify when something is out of standard—meaning that a quick look at the area should determine whether it is within standard, and anything out of standard is quickly identified. Visual controls can take many different forms, depending on the setting. The following are examples of effective visual controls:

1. Shadowboard—An example of a shadowboard is shown in Figure 8.1. Shadowboards can quickly identify when an item is missing or was not returned properly. These boards are just as they sound, as a dark image of the tool is placed under the location where the tool is stored. If you can see the "shadow," then you know that the tool is not there; if the tool does not match the shadow, then you can see that the tool is placed in the wrong location. A famous example of shadowboards is the Alcatraz prison kitchen, which had shadows behind each kitchen knife. Although identifying the tool in your facility is likely not a life-and-death situation, quickly identifying when an item is missing is highly important.

2. Taping/painting/marking off location—An example of this is shown in Figure 5.6, where plastic carts are being placed in their defined location. This is often used for marking off the specific location of an item on the floor. Storage racks, items used for conveyance, or work areas are examples of items that can be marked this way. These items should either have a set location or a home location where items should be returned. This is similar to a shadowboard, but is used for items other than tools.

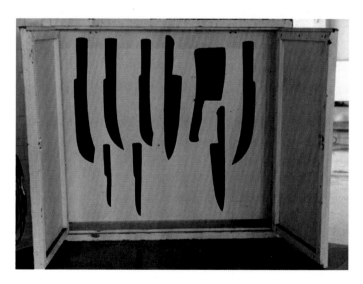

FIGURE 8.1
Shadowboard used at Alcatraz prison to make sure that all knives were accounted for.

3. Mark gauges, equipment, or areas with acceptable range—Again, the intent of these visuals is to easily see when something is out of standard, so a min max on inventory, gauges, equipment, etc., creates a quick visual cue that allows anyone to quickly know if there is something out of the ordinary.

4. Whiteboards—The first three items show when a process is out of standard. Whiteboards can be an effective visual to show when other items are not to standard. An example of a whiteboard visual that we have implemented is showing employee assignments for different lines, and listing the standard number of employees for that line that day. Not only does this provide the standard of who should be working on a given line each day, but adding the names and comparing that to the standard labor can quickly identify if too many or too few employees are assigned to that line.

If the process passes the point test—meaning that the "how" can be explained by pointing at a visual control and offering a brief explanation—then the visual control is effective and the visual can be made permanent. However, if circumstances warrant keeping the temporary labeling for an extended period, then the visual should proceed back through the PDCA process until it can meet the aforementioned requirement. Remember the dry ingredient carts in Figure 5.6, where we initially used yellow tape on the floor to mark the lanes for each cart. After pallet jacks started to rip up the tape, we moved the tape to the wall. Although we discussed painting these lane lines on the wall, we made the decision that because the tape was staying on the wall, we would keep only the tape for 6 months. The timing allowed us to ensure that the layout did not need to be changed, and that there were not any plans to alter the area in the near future. Although painting over these lines would be a quick fix, we also needed the continued buy-in from our designated facilities maintenance employee, so we wanted to ensure he was not doing double work when we could.

STANDARDIZED WORK

Document the final standardized work for each process that you implemented. Using the process flow that you drafted in watching the work, you will now begin to formally draft the standardized work sheet for each process listed in the process flow. To review the steps that got you to this point:

1. Watch the work.
2. Document the steps of each process—the what, how, and why, and analyze.
3. Regroup with team, discuss ideas, get input, and make action items against ideas.
4. Repeat steps 1–3 for all work processes.
5. Implement a pilot—do PDCA.
6. **Implement the final process**.

Having gone through steps 1–5 at least once, and possibly more by going through the PDCA cycle, the process should be well documented in some form already. In this step, you are memorializing the current standardized work practice. Although this standard can still be improved upon, the current standard must be documented and followed while it is the standard.

In drafting the what, how, and why for a process, take it one step at a time and use common sense on breaking down each step. If a process involves pushing clips to join two different items, although the process is completed in one fluid motion, it may make more sense to document and eventually train this process as a few different steps. This process might look like this:

Step 1

- What—grab clips
- How—using left hand, grab small handful of clips from orange bucket
- Why—you will feed the clips to your right hand for installation

Step 2

- What—install clips
- How—starting at far right, feed clips from left hand to right hand and install in all "A" slots
- Why—to secure part

Although the process might look like one fluid motion, break it down to the fundamentals. In this example, grabbing a small handful of clips and installing the clips appear to be one fluid process. However, learning how to do each action individually is important to complete the overall process. Think of learning something new yourself. If you have ever played golf, you know that there is much more to hitting the ball than just hitting the ball. You need to use the correct club, be lined up correctly, and swing appropriately. Each of these steps is distinct, even though "hitting a golf ball" appears to be one process. The standardized work sheet is the time to get into the details of the process; it will be the document used to audit the process, to train the process, and to analyze when improving the process, so it must make sense and have the details.

As shown in the preceding example, the "what" element of the process should be short, and just describe the basic step. In the golf example, the "whats" might be select club, line up, and swing club. The "what" just gives the basics that the "how" will explain. The "how" for a particular step in the process may have multiple steps itself, and oftentimes does. In the "how," you are explaining the step-by-step process of completing the "what," so provide details on how every part of the process needs to be performed. For every "how," give the reason why it needs to be performed. The "why" not only provides the reason behind each part of the process, but when workers know why they are performing a task, they are more equipped to improve upon it. The "why" gives them their stated objective for the process, so it allows them to use their creativity to think through other ways that the goal may be accomplished.

In addition to listing each step of the process, pictures of key points—or even a layout of an area showing where items can be retrieved—are highly beneficial. Someone who does not regularly perform a process who is asked to fill in may need to refresh themselves on the steps; in such cases, the presence of written directions with visuals enables a quick under-standing of the process. Also, anyone can pull up the standardized work to gain a better understanding or better audit a process if it includes the visual key points. Pictures allow a person to read through the standard-ized work elements and have the corresponding picture to remind them of, "Oh yeah, I remember how to do this." Shop floor workers do not run into this scenario as much as office workers (who usually have a "backup" for when they are out for vacation or otherwise, with minimal training on the different tasks) do. They may have been shown how to perform the steps of the process once or twice, so having the pictures facilitates remembering everything involved.

Post all the standardized work sheets together in a central area where they can easily be accessed. Employees should not need to reference these documents during the workday if they have been properly trained, but they may occasionally need to quickly scan over a document if they are required to perform a process that they have only been trained on and never performed during a shift. The true purpose of posting the standardized work sheets in a central area is to facilitate auditing, which is explained in Section III of this book.

ASSESS YOUR GOAL

Compare the area with the goal for each of the 5Ss outlined previously, as well as the standardized work goals. Processes can continue to be improved upon, but the 5S goals need to be reached before you move to the next area. Do a self-evaluation with your team on whether the 5S elements are complete. Also have someone in management come in and sign off on the 5S; your team obviously will be partial to the work that has been done and be too close to the work to realize that more work may need to be done. It is the same concept as bringing in outsiders to assess the quality of visual controls and whether the documented steps of the standardized work are specific.

Once Process Implementation is complete for the entire area, train other employees in the company about Process Implementation and walk them through the completed area. This allows others to see firsthand the results of a Process Implementation and also learn how proper Process Implementation should look.

Having a completed Process Implementation will add to the reflection phase for a different area, as the completed area can become a place for others to watch processes and gather ideas. Once the processes are in place to sustain the implementation, others in the company can again look at your area to learn about sustaining processes.

Section III

Sustain

Just as Section II ("Planning") took up a substantial part of this book, Section III ("Sustain") also takes up a large section. The amount of information in these two parts shows their importance in the Process Implementation process.

9

Auditing

Auditing maintains your strong foundation of standardized work and visual controls, and thus prevents setbacks in the work that you have done to date. Without auditing and sustaining all the work that you have put in thus far, the area will easily fall back into the state that it was in before you began. The benefits of auditing are:

1. Gives employees an avenue to improve standardized work
2. Makes problems visual
3. Allows tracking of issues so no issue falls through the cracks
4. Provides a means of communication between shifts and allows for problem solving to occur between shifts ("you're seeing that to, here's what we have tried," etc.)
5. Records issues, which allows for more effective problem solving, and also heightens accountability
6. Maintains the consistency of the process, and thus the output of the process

Both the standardized work and the 5S components of the area should be audited, the frequency of which should be realistic and negotiable (once or twice daily is all that is needed per shift). The following is a sample plan for auditing both 5S and standardized work:

a. Team leaders—two daily
b. Supervisors—two weekly
c. Manager—one weekly

In addition to these audits, the role of leadership is obviously to stop an employee who is performing a process out of standard and immediately correct them. The audits are designed to be a set aside task that ensures an actual audit of the process is completed, which will be more in-depth than noticing someone doing something incorrectly. An audit should be done when the initial intent is performing the audit; do not count the process as being audited just because you happened to be having a conversation with the employee while they performed a process.

Doing an audit allows you to coach the operators and/or leadership to drive continuous improvement in the process. The employee involvement throughout the planning phase will create employee pride in the area and create a sense of self-ownership, which will drive sustainment efforts by the employees. In an ideal state, workers will convey potential improvement ideas for performing a process when you are performing your audit, as they both understand that while you are ensuring conformance to the standard, the audit allows for an analysis of the work, and they understand their role as being responsible for driving this improvement. Improvement is accomplished one step at a time, and auditing prevents you from falling back a step once you have made an improvement. Figure 9.1 shows this incremental improvement that is reliant on auditing to sustain itself at each step.

A major component of employees driving improvement involves tracking issues and countermeasures. Each workstation should be set up with a Countermeasure Sheet, which is basically a table with columns for

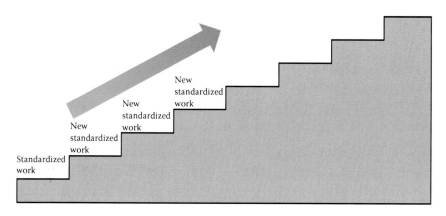

FIGURE 9.1

After establishing standardized work, each small improvement step must have new standardized work in order to maintain the improvement. To maintain this improvement, effective auditing must be performed or you will be descending back down the steps, rather than ascending with continuous improvements to reach your ideal state.

the following data: date and team member, the issue, the initial action/countermeasure taken, comments, and "reviewed by" initials. An example of a section of a Countermeasure Sheet is shown in Figure 9.2, but this is a basic Excel document that can be easily created. As shown in the five "why" examples from Taiichi Ohno, the goal in Lean is to identify the root cause and implement a countermeasure that will solve the root cause issue. In auditing, determining the root cause and implementing a countermeasure may not be possible, but an initial step can be taken by the employee or supervisor to keep the process running. Some countermeasures may need specific individuals or teams of individuals assigned to them, with a target date and plan to address the issue, whereas other countermeasures may be a quick fix.

The role of the supervisors is to monitor the Countermeasure Sheets throughout the shift and make sure that each issue has a corresponding countermeasure to ensure the health of the auditing system. Supervisors should take immediate action(s) on countermeasures when applicable, and then initial the Countermeasure sheet in the "Reviewed By" column. If the Countermeasure requires the assistance of another department, then log that information in the comments section. For example, if a work order for maintenance must be submitted to clear the issue, then put the work order number and expected completion date in the comments section so that follow-up can be done easily. If the employee is able to take action to immediately help address the issue, then it should be encouraged that they do so. Having the employees work to solve problems not only frees the supervisor to concentrate on other tasks, but it also challenges the employees to solve the problem and further engages them in the improvement of their process.

Countermeasure sheet					
				Completion date	
Date	Issue	Action required	Who	Target	Actual

FIGURE 9.2
Countermeasure Sheet used to track issues and countermeasures.

In summary, the process of utilizing the Countermeasure Sheet involves the following:

1. ID the issue—the employee should be identifying the issues, but anyone from the shop floor through upper management may identify an issue.
2. Log the issue and the initial action taken—on the Countermeasure Sheet, the issue should be logged, as well as the initial action taken. Initial action taken could be simply notifying the supervisor, notifying maintenance, etc. However, if an issue is identified, then an initial action MUST be taken.
3. Supervisor checks the Countermeasure Sheet throughout the shift, initials any new items as being acknowledged, and follows up with the employee to gain more information on the issue.
   a. If the initial step taken solves the root cause of the issue, then the item is closed out.
4. If resources are needed.
   a. Log in the comments section what is required, by whom, and the expected completion date (e.g., Work order 91322 submitted to Maintenance on 6/11/15 for installation of X, to be completed by 6/22/15).
   b. If operations conducts a production meeting where such items are addressed, then take the countermeasure to the production meeting for assignment to the proper department and the target completion date.
5. Once completed, put in any follow-up information for the issue and fill in the completion date.

Figure 9.3 shows how this auditing process fits into the overall implementation process. The checking and adjusting that are done during auditing and countermeasures sustain all the work that was done to develop, implement, and train.

In auditing, the supervisor's role initially is to choose the operator or process to audit. Each process should be audited at least monthly, so ensure that each operator is being audited. To determine which process to audit, judgment should dictate where to audit by which process needs the most attention. For example, focus on employees who traditionally do not find any issues. Perform the standardized work audit below, the 5S audit, or the T-Card audit (either using actual T-Cards or the audits outlined

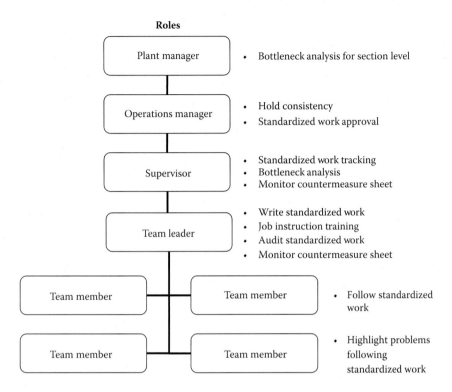

Roles

- Plant manager
  - Bottleneck analysis for section level
- Operations manager
  - Hold consistency
  - Standardized work approval
- Supervisor
  - Standardized work tracking
  - Bottleneck analysis
  - Monitor countermeasure sheet
- Team leader
  - Write standardized work
  - Job instruction training
  - Audit standardized work
  - Monitor countermeasure sheet
- Team member / Team member
  - Follow standardized work
- Team member / Team member
  - Highlight problems following standardized work

FIGURE 9.3
The structure of your organization may look different, but roles should be identified at each level, with team leaders writing and training standardized work, and team members following and finding improvement opportunities.

in that section). The role of the supervisor once they have chosen what to audit is to

1. Coach/Mentor—pass along their knowledge to the employee
2. Log that the audit has been completed and update the audit tracking sheet
3. Report out audit findings

A high-level overview of each level's role in developing and sustaining standardized work and visual controls is shown in Figure 9.3. Your company's structure may have different titles than those listed, but ensure that the tasks given for each role are dispersed across the responsible positions. The responsibilities outlined in Figure 9.3 are only for developing and sustaining standardized work, as I'm sure upper level management has more pressing responsibilities during the course of the day than tackling bottleneck analysis and approving standardized work.

When I was a supervisor, one of our employees noted that they were missing their broom on a couple of different days. The employee noted the issue with the countermeasure that they notified their supervisor, myself (which was the correct course, as the employee should not go searching for an item when they have their actual work to complete). The first time that I was notified, I found the broom in the warehouse area, returned it to its location, and notified the warehouse supervisor that the broom had been used and not returned. I then updated the Countermeasure Sheet to reflect that the broom was found in the warehouse and that the warehouse supervisor was notified. After this happened a second time, the conversation with the warehouse supervisor was more in-depth, and I suggested that he talk to the appropriate person about purchasing a broom to keep at the location where I found our broom both times, and that in the meantime, if a broom was borrowed, it must be returned to the location from where it was taken. We set a date for when the broom for the warehouse would be installed, and I noted this on our Countermeasure Sheet.

Although having a broom seems like a small issue, the broom had a set standard location, and when it was not in its location, the employee had to know that the supervisor would ensure the area was kept to standard, and that they had the necessary tools to complete their work. In essence, not only did the employee need the broom to perform their job, but taking action immediately to locate this item and follow up with who had taken it showed the employee that we were serious about keeping things to the standard. Figure 9.4 shows the broom station, and illustrates how it was easily evident for the employee to notice the missing broom.

In some instances, the employee will be able to take action without the need of a supervisor, and it is important that this information is logged so that management can review the issue and countermeasure, and initial that the issue has been acknowledged. Writing these down also ensures that the issue gets communicated to everyone, and management can follow up with the employee to gather more information about the issue. The issue may be small, but it is recurring and needs resources to fix. Logging any issue also communicates issues between shifts, and all shifts can work together on finding a solution.

So far, we have addressed the role of the employee and supervisor in auditing; now, we will discuss the role of management. These different roles

FIGURE 9.4
Broom station outside worker's room.

can be broken down to differing management levels depending on the structure of your organization, but basically the management's role is to

a. Audit the health of the system and ensure that roles and responsibilities have been clearly given.
b. Know what the subordinate is working on in relation to the audit system, and how did they deduce that this is what they should work on.
c. Mentor/coach—enable your employees to drive improvement by giving them the knowledge necessary to make decisions—generally from the manager level to the supervisor level.
d. Ensure that the subordinate is complying with the standard for frequency (i.e., they are performing the number of audits that they are required to perform) and also that they are finding issues on their audits and properly addressing them.
e. Look for consistent issues with a process or an employee.

It is critical for management to play a role in the auditing process to show the importance of the process and to work with supervisors to improve the area and the auditing process.

STANDARDIZED WORK

For standardized work, create an audit schedule and watch the employees do the standardized work to ensure that they are following the steps of the process. The purpose of standardized work audits are as follows:

1. It maintains control over the process—Adhering to the standard and performing regular audits ensures that it is continually followed and that it does not get out of control.
2. It maintains a disciplined Lean environment that follows standards and strives for improvement.
   a. I find that traffic lights that only contain the light for the turning arrow to be somewhat unnecessary. Many cities have worked around this by making the turning arrow flash yellow when the straight traffic has the green light. Although this now seems to be the norm, I still run across traffic lights where the left turn lane has its own light that stays red when the straight traffic is green. Although I know that the better option and generally accepted option is to have a flashing yellow light, I abide by the traffic light because I know that if I turn on red, I might get pulled over. Employees need to have this same feeling when doing something that is not the standard (and work to get the standard changed).
3. It allows for process improvement. Follow the Process Implementation process again; each time you watch a process, take advantage of the opportunity for improvement and follow the Process Implementation cycle of watching the work, talking to the employee, analyzing, and reflecting, to determine if there is a better way to perform the process.
4. It enables tracking of issues and countermeasures—Any time that someone, from shop floor worker up to upper management, finds an issue, both the issue and countermeasure that was put in place should be logged on a sheet. The format of the sheet should be simple, and everyone should use the same sheet. A sample Countermeasure Sheet is shown in Figure 9.2.
5. Auditing ensures adherence to the standard—If an employee is deviating from the standard, as with training, follow up immediately on the correct way to perform the process, and gain an understanding of why the employee was not performing the process to standard.

One way to effectively audit is to take the standardized work sheet that you created or is being used to document the steps of the process, remove the Why section, and replace it with a grading system. Mark in this new column whether the step was followed as written, whether the step was completed or missed, and any observations you have about watching the process. This exercise is very similar to the planning exercise of watching the worker perform the process and taking notes on things that you see. This practice not only ensures that workers are performing the standardized work to standard, but it also drives continuous improvement by allowing you to go to the Gemba and again watch the work. It is the repetition of this PDCA that sustains your results and continues to drive improvement; auditing is checking and adjusting the process that was implemented during the initial PDCA.

Related to coaching, these audits are not intended to be a "gotcha" exercise. As stated previously, at some point, coaching must cease and discipline must take over, but ensure that enough coaching has taken place, and that the employee has the correct knowledge needed to perform the process before resorting to discipline. (If you need to continually try to catch employees not performing the process the way they have been told, you likely have a much bigger issue that must be addressed before any true Lean can be established.) The fundamental purpose of performing an audit is to improve standardized work, so as with everything else that you have done thus far, continue to be transparent and make a point of explaining to the employees the purpose of performing an audit.

In building a relationship of trust, do not attempt to hide from the employees when performing an audit. The employee should know that the expectation is to follow the standard, and that you are watching them perform the process to ensure compliance, but also to gain any potential improvements. If the workers have begun to stray from the standard in their everyday performance of the process, it will be evident from the audit, as their habits will have changed in how they perform each step. As with performing the initial Process Implementation, follow the same approach for auditing standardized work. Watch the work, ensure that the work is being performed to standard, analyze the work with input from the team member, reflect on the work, and go through the planning cycle on thinking through any potential improvements.

5S AUDITS

5S audits should focus on visual controls, with the visual control showing when something is out of standard. The same rule applies for 5S audits as with standardized work audits; reinforce that they are not done to catch employees doing wrong, but rather, they are intended to sustain your implemented processes, coach your employees on doing the processes correctly, and make the issues in the area visual. To sustain the implemented processes, standards must be created for workers. Standards must answer the following: who, what, when, where, and how.

The different elements of 5S can be audited in different ways, of which several examples are discussed in the following list:

1. For Sort, have employees check their area for unneeded items. Without having to sign off on an audit sheet, employees will likely either deal with having the unneeded item in their area, or will move the item out of their area, but forget to tell a supervisor that this task had to be done. Having employees mark off on an audit sheet reminds them to check for unneeded items, prompts them to write down if any unneeded items had to be removed, and creates a visual to see that an unneeded item was in the area on a certain date and a certain shift. Having this information allows you to walk the issue back and prevent the unneeded item from appearing in the work area again.

2. For Set in Order, have employees audit their area for the right tools and material, in the right amount. Employees having to look for material wastes time, but employees will hunt down what they need because they have to do their jobs. The audit should prompt employees to check that they had the right tools and material in the right amount at the start of their shift, and that they left the area in this condition at the end of their shift. Auditing Set in Order highlights waste of employee movement in searching for tools and material. It also highlights waste of overproduction, and tracks a potential cause of downtime.

3. For Shine, the workers should audit the condition of equipment at the start of shift, a cleaning schedule should be established, and workers should audit their area for spills and cleanliness. Auditing the condition of the equipment at startup and shutdown is vital to prevent downtime at startup. If workers in the previous shift sign

that they left the item in good condition for the following shift, and their counterparts in the following shift audit the item before they begin their shift, then any potential issue is made known before the day's operations begin. The importance of establishing a cleaning schedule on the audit ensures that everything in the area is getting cleaned on a frequent basis. Auditing the area for spills and cleanliness allows the worker to track the occurrence of an unclean condition, which shows that the cause of the spills needs to be addressed.

Accomplishing an audit for each of these can be done in a few different ways:

1. A plotter map—Having a blueprint of the area that outlines where different checks should occur and what should be found at each check station (e.g., trash can, broom, certain tools) provides a visual cue to what needs to be checked.
2. A walkthrough of the area can be performed to determine if items are out of standard. If visual controls were properly implemented, performing a walkthrough of the area to determine when something is out of standard based on the visual control should be adequate. Just make sure that the visual control matches what is actually occurring, and note anything out of standard.
3. Using the 5S assessment scorecard to score the area—The 5S assessment that was used in determining your current state and setting the goal at the beginning of the Process Implementation can be used when auditing the area, to again determine the current level of each item.
4. Create audit sheets for the workers in the area to sign off for each of the first 3Ss. Include in the audit sheet the key elements for each "S." Supervisors should be auditing the sheet each day to ensure that the employees are doing the audits and assisting in establishing countermeasures for items out of standard.

As previously stated, reward systems help sustain the 5S condition. This system could be as simple as auditing all the areas in the company each month and giving a gift card to the employees in the area that had the highest scores for their 5S. If there is a bonus structure for different key performance indicators, you can include maintaining the 5S and standardized work at a certain level to receive a portion of the bonus. A reward system is not essential to sustainment, but it will provide a great assist mechanism.

10

T-Cards

Another book could be written on the T-Card process, but this section will provide you with an overview of how to successfully implement a T-Card system into any work area. A T-Card system is an effective way to

1. Provide visuals to your employees to sustain their area.
2. Hold people more accountable to perform their tasks.
3. Give management a quick visual indication to see if a process has been completed and if there is an issue.
4. Improve standardized work.

T-Cards are another means of auditing, and they are very effective because of the visual controls that they provide. Not only can T-Cards be used for standardized work and 5S audits, but they can also be a great way to ensure proper startup and shutdown. Remember in Shine that workers were to leave the tools in ready-to-use condition, and employees at the start of the shift are also supposed to inspect the tools for ready-to-use condition. Having a T-Card system establishes a system to remind the employee to perform this check, and a visual for the supervisor to see if the check has been performed. This is just one example of the potential use for the T-Card system; other examples include critical daily tasks, infrequent tasks, or checking the area for the proper setup.

Rather than have audit sheets for employees and management to check certain items, a T-Card system can be put in place to perform the same objective. A T-Card is as it sounds, it is a card shaped in the letter "T." The top of the T, the horizontal section, will have the name of the process that should be completed, or the general name of what the employee is checking (e.g., startup audit check on machine, morning responsibilities, list of tools to check). The body of the T, the vertical section, will give the employee the list of duties to perform. The front part of the card, the part that lists the employee responsibilities, should be green, and the back of the card should be red. An example of a T-Card, showing both sides, is shown in Figure 10.1.

T-Cards fit nicely into time card holders, which are easily mounted on the wall. When employees begin their shift, the T-Card should be placed in the holder marked incomplete with the green side facing out. This is a visual for employees that notifies them to complete the process, and is also a visual in auditing, as it shows that the process is not complete. Once

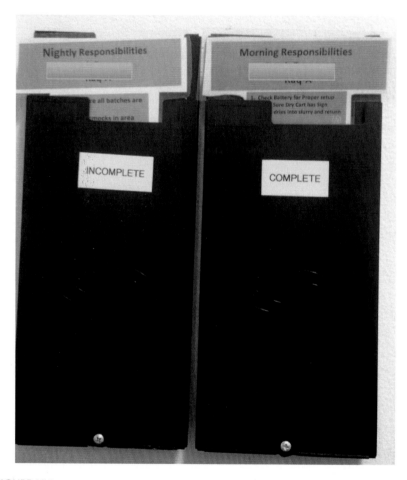

FIGURE 10.1

Basic T-Card setup, an incomplete and a complete rack. The morning responsibilities have been completed without issues, and thus the green side is facing out. The nightly responsibilities still need to be completed.

complete, the employee moves the T-Card to the holder marked complete. If there are no issues in completing the process, the employee keeps the green side facing out. However, if there are issues, the employee flips the card so that the red side is facing out. If red shows, the employee should mark why the T-Card was red on an action item list under the T-Card, and what action was taken in response to this being red, the same process as discussed earlier for the Countermeasure Sheet. When auditing the area, the red of the card becomes a quick visual that shows something is out of standard in the area. The advantage of this system is that it makes problems more visual, as you can quickly look in a large area and see whether anything is red, without having to go look at each audit sheet in the area and read whether anything was out of standard. Another advantage is that employees can walk their area with the card, and reference the card to make sure they are completing all the tasks.

Having a T-Card alone will not sustain the area, but auditing the tasks on the T-Card will. In this system, employees do not need to sign off on performing the tasks, as the card being moved to complete shows that the tasks have been done. As with the 5S audits, supervisors should still audit the T-Card completeness, make sure that a red T-Card has an action item, and follow up on any issues listed in the action item list. Also, they should maintain records of the completed action item lists to monitor repeat issues and the countermeasures taken.

There are five types of audits than can be conducted using the T-Card process:

1. Job Knowledge
2. Coaching
3. Process Compliance
4. Follow-up Process Compliance
5. T-Card Health

A variation of any of these audits can be used if you do not have a T-Card system in place; if you wish to audit standardized work without a T-Card system, follow the steps of any of these audits and do them without T-Cards.

1. Job Knowledge Audit—This is an audit of employee knowledge of process, generally done at the beginning of implementation to ensure that employee knowledge matches the expected level of expertise for

the process. It can be used as a follow-up as well. To perform this audit, have the employee perform the process as you observe, correcting any step that is not done correctly or is missed.

2. Coaching Audit—Take a T-Card that has been moved to completed and is showing green. Walk through the process with the employee and ask them questions about the process (e.g., Why do you do this, What would you have done if…). Point out things that the employee may have missed, but do so in a constructive manner. Use this opportunity to ask the employee about any ways that the process can be improved.

3. Process Compliance—Identify a process in which the employee rarely, if ever, identifies a problem. After the employee has moved the card to completed and kept green, go through the process and try to find anything that is out of standard. If anything is found out of standard, walk the employee through the process, showing them the abnormalities, and explain the importance of identifying these out-of-standard conditions. Employees sometimes need a reminder of the importance of each task that they are doing, and this type of audit can provide that.

4. Follow-up Process Compliance—Exactly as it sounds, this type of audit enables you to follow up with the employee that you audited in a Process Compliance Audit and make sure that they are now finding the problems that they should.

5. T-Card Health Audit—For management and supervisor to work on improving the T-Card station. Although called a T-Card Health Audit, this can be done for 5S and standardized work audits described earlier. What should be addressed are issues related to the following:
   a. Are issues being found on all shifts?
   b. Are "initial steps taken" being filled in for all issues?
   c. Are target dates identified?
   d. Is management keeping to their standard check (e.g., weekly, monthly)?

Again, any of these audits can be done without having a T-Card system in place, but the actual T-Card will provide an easy visual, and instills more accountability into the employees as moving the card to complete means the employee did the process in its entirety, and did not find any issue when doing the process.

11

Training Standardized Work

It is the responsibility of the company to teach the employee how to perform their tasks. The saying in Lean is that, "if the student hasn't learned, then the teacher hasn't taught." Without properly training the employees to perform the task, the implementation and sustainment efforts are meaningless. Drafting nice Standardized work sheets and posting them everywhere is nice, but not ensuring that a system is in place to adequately train the employees on the standardized work makes these written documents valueless.

Up to this point, we have discussed empowering employees and building trust through involvement with the more seasoned employees on the shop floor. However, the process of empowerment and trust actually begins once employees are hired and begin training.* Not only are workers more likely to have less stress and not look for other employment when they have the required job knowledge, but leaders bolster trust when they share the information that workers need to perform their jobs well (Mayfield and Mayfield, 2002, p. 91), and this information empowers their employees. Furthermore, workers are able to improve processes when they understand why each step is being performed.

Empowerment involves preparing your staff by boosting their confidence and competence, and by communicating a clear vision and goals (Williams, 2014). Effective empowerment of your staff can lead to higher levels of employee motivation and satisfaction, lower levels of stress for employees, higher levels of employee skill development, and better time

---

* Having standardized work is a prerequisite for training; you cannot train employees on how to perform a task if that task is not yet standardized, and thus is why training has followed implementation in this book. If you are able to perform offline training when new employees are hired, a dedicated trainer or supervisor should lead the training. If you are not able to provide this type of exclusive training, a shop floor employee that is trainer certified should provide the training. Levels of job skill are discussed in the "Multi-Function Worker Chart" section.

management for managers (Williams, 2015). As with other principles outlined in this book, proper training also increases trust, job satisfaction, and employee motivation.

Training and managing employees draws many similarities to an athletic coach, as each must teach their personnel the proper techniques and then expand those skills into plays or processes, build trust with personnel and have effective communication, and hold personnel accountable for not following the trained process. Throughout this book, we have been talking about the importance of coaching, and training is one of the most vital areas. A supervisor may be a disciplinarian if necessary, but the primary role is instructional, always leading a small "learning group" of workers (Huntzinger, 2006). According to Vince Lombardi, if you are coaching then you are teaching, as he once said, "They call it coaching, but it is teaching. You do not just tell them...you show them the reasons." Vince Lombardi understood the importance of explaining the What, How, and Why to his players, and the results speak for themselves, as he won five National Football League Championships and two Super Bowls (not to mention having the Super Bowl Trophy bear his name).

Standardized work and training your people go hand in hand and is part of the overall system (Liker and Meier, 2007). Companies perform training in myriad ways; most seem to have an orientation period that explains the background of the company, and then may extend into a longer training period (a couple of days to a week). When I worked at Toyota, I started with a large group of new hires that spent 2 weeks having classroom training on the Toyota Production System (TPS), which also incorporated some physical training led by trainers from Toyota's fitness facility. After two weeks of in-class training, each "team member" was then to report to their assigned group where they would be working. I was assigned to bumper paint, and spent another week with four other new team members working on the first shift (I continued to work on the first shift, but three of the other four team members would work on the second shift after the training; the training Team Leader at Toyota is a first-shift position so employees' first week of work is made to accommodate a week of training). During this week of training, each of us was taught not only each position in the group using the Job Instruction (JI) method, but we were also taught and had to reach a certain goal for ancillary tasks such as screwing screws or nuts, which involved trying to pick up a certain number each time in your hand and then working to rotate them in your left hand while screwing them in with your right with the use of a drill gun.

After each training session, we had to pass a test that required doing a certain number of tasks in a given time before we could proceed to learning the next process. In addition to classroom training on each of the 16 jobs in the group, each of us also worked with the workers performing these jobs in the group during actual production for short periods. Upon completion of this classroom training and official implementation into our groups, we spent another week working side by side with an experienced team member who was trainer level (more on this later when we talk about Multi-Function Worker Sheets), slowly working our way up from performing only part of each task to performing 100% of the process each time.

Thus, the training I had at Toyota consisted of the following:

- Two weeks in class training with physical training (with additional classroom training throughout employment)
- One week of Job Instruction Training by Team Leader, performing each task for the Team Leader, and working with Team Members currently performing the processes online
- One week of incremental performance, working with trainer level Team Member and performing a percentage of task each day, working up to 100%

Four weeks of training each new Team Member is a luxury that most companies cannot afford, and time dedicated to training is not something most already overloaded supervisors can accommodate. However, if the principles of the JI explained in the following discussion are appropriately put into practice, your employees can be set up for success in a shorter amount of time.

U.S. General George Patton once said, "Don't tell people how to do things, tell them what to do and let them surprise you with the results." Patton could make this statement because of the effective training program implemented by the United States Army during World War II, a training program that continues to be used today by companies such as Toyota to train new employees. The Job Instruction Program is a four-step process designed to help supervisors train on new jobs and involves putting the worker at ease, and then walking them through the steps of the task and the reasons for each step. After explaining the task to the worker, the worker then performs the task and receives immediate feedback and correction. The JI method was originally on a two-sided card that supervisors carried with them; this two-sided card is shown in Figure 11.1.

HOW TO GET READY
TO INSTRUCT

Have a Time Table—
how much skill you expect him to have, by what date.

Break Down the Job—
list important steps.
pick out the key points. (Safety is always a key point.)

Have Everything Ready—
the right equipment, materials, and supplies.

Have the Workplace Properly Arranged—
just as the worker will be expected to keep it.

Job Instruction Training

TRAINING WITHIN INDUSTRY
Bureau of Training
War Manpower Commission

KEEP THIS CARD HANDY

GPO 16—35140-1

Front of the Job Instruction Card

HOW TO INSTRUCT

Step 1—Prepare the Worker
Put him at ease.
State the job and find out what he already knows about it.
Get him interested in learning job.
Place in correct position.

Step 2—Present the Operation
Tell, show, and illustrate one IMPORTANT STEP at a time.
Stress each KEY POINT.
Instruct clearly, completely, and patiently, but no more than he can master.

Step 3—Try Out Performance
Have him do the job—correct errors.
Have him explain each KEY POINT to you as he does the job again.
Make sure he understands.
Continue until YOU know HE knows.

Step 4—Follow Up
Put him on his own. Designate to whom he goes for help.
Check frequently. Encourage questions.
Taper off extra coaching and close follow-up.

16—35140-1

If Worker Hasn't Learned, the Instructor Hasn't Taught

Back of the Job Instruction Card

FIGURE 11.1
The two sides of the Job Instruction Card show the preparations needed to be made by the trainer prior to instructing, and then provides a detailed step by step process of how to instruct.

As you can see, the front of the card ("How to Get Ready to Instruct") is intended for the trainer, and provides a list of items that should be prepared in advance of the training.

1. Have a timetable of how much skill you expect the worker to have, and by what date. It is important to set goals for the employees—both for them to know when they are expected to know the task, and for the instructor to keep track of how much knowledge the employee should have by what date.
   a. Specific goals are more effective than vague and general goals.
   b. Difficult, challenging goals are more effective than relatively easy and common goals. Goals must be reachable though.
   c. Accepted goals set in participation are to be preferred over assigned goals (this will be discussed in more detail in the Multi-Function Worker Chart [MFWC] section).
   d. Objective feedback about the advances attained in respect to the goal is absolutely necessary but is not a sufficient condition for the successful implementation of goal setting (Salvendy, 2012).

Goal setting offers a useful and important method to motivate employees and junior managers to achieve their goals (Salvendy, 2012). Although this sounds elementary, people want to know exactly what you expect from them and by what date. I played football in high school; if my football coach had said he wanted me to be stronger in 6 months, I would be a little confused and think to myself that I will obviously be stronger, but wonder how much stronger he will expect me to be. However, if he said that he wanted me to increase the amount I can lift by X on each different exercise, then I would have a goal to work toward, rather than a vague request with the stress that what I accomplish might not be what he had in mind. As crazy as it might sound, the vast majority of people, including myself, would work harder to attain the specific goal of lifting a certain weight by a set date, rather than work extremely hard to reach the generic goal of getting stronger. This is probably why my football coach set specific attainable goals for each person for how much weight we were expected to lift and by what date; he's a smart guy and he understood this concept.

2. Break out the job with important steps and key points—Giving the worker an overview of the process before going through each step in the training will provide a preliminary understanding that allows the worker to see the completed process before going through each step. Sticking with the football example, when learning a new offensive play, the most effective process is to go through the entire play showing each player's role and the goal of the play. Once everyone sees the big picture, they can learn their specific task, knowing "why" they are performing the task that they are performing (the goal).

3. Have Everything Ready and the Workplace properly arranged—Having everything ready and having the workplace properly arranged are somewhat self-explanatory. The goal is to have the worker be able to perform the task, and you therefore want to have the materials that they will be using and the workplace properly arranged as it will be once they are "on the floor." I've had the opportunity to work closely with the U.S. Military through working for a company that provides military rations. Even in training, the military operates with a "train as you fight" mentality—meaning, not only are they doing training in the field to simulate battle, using the exact equipment that will be used in battle, but they are also eating exactly as they will in battle. This is the same principle as training your

employees; training is the one time that you have set aside to provide the fundamentals, and you should make the conditions during training match those that will exist on the shop floor. If we think about the football example of learning a new offensive play, although the offense will usually learn the play before executing the play against a defense, before the offense runs the play in a game (i.e., before an employee performs the job on the shop floor), the offense would be lined up against the expected defense when going through the steps of the new play, so that the scenario would be as close to real as possible for a game setting.

The information found on the back of the card explains the process of breaking down the job into different steps. To properly break down the job, you will need a document that explains each step and key points. If you went through the process of gaining stability, then you created a Standardized work sheet, listing the process out in a "what, how, and why" format. Two major components of getting subordinates ready to have tasks delegated to them are boosting their knowledge and skill levels and helping them feel competent (Williams, 2015). Some keys to effective teaching include:

a. Telling them what they will be learning and why
b. Providing information and demonstrations
c. Allowing opportunities for practice
d. Providing feedback on performance (Williams, 2015)

The framework provided for training on the Job Instruction Card has withstood the test of time and continues to provide a very effective method to teach employees to learn a new process.

Some training programs recommend having a separate training document, a JI document that identifies the steps of the job, what is important about each step, and how to perform each step. Explaining what the step is, how the step is performed, and why the steps are being performed accomplishes the same goal, and prevents completing an entirely new document on which to train (which must also be edited each time the standardized work is changed). With your Standardized work sheet, you are now ready to train.

Training doesn't cease once new hires are trained for the shop floor, but is a continual activity for all employees. When new processes are implemented or improved, the same training regiment should apply. Training current employees on new or improved process, or providing training to employees on items of relevance for the furtherance of their careers builds trust and empowers employees the same way as training new hires and improving processes through Process Implementation.

HOW TO INSTRUCT

Remember back to Taylorism, where the process was dictated using the scientific method, meaning that supervisors determined that there was one best way to perform the job and any deviation from this process was unacceptable. In this training, not only are you ensuring that the employee becomes comfortable with performing the task, but in explaining the key steps, you are laying the foundation for improvement, as the employee cannot improve what they do not understand.

The JI training can be accomplished in a group, but with no more than four to five employees. As with other steps in the Process Implementation, the JI Training follows a Plan–Do–Check–Adjust (PDCA) cycle, which is shown in Figure 11.2.

After having done the initial work to plan the timetable for the training and prepare the area for training, you are now ready to perform the steps on the back of the card, which provide the trainer with directions on how to instruct. The following is a detailed explanation of each of the four steps found on the back of the Job Instruction Card.

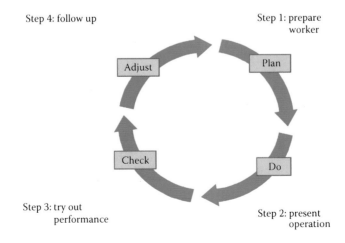

FIGURE 11.2
Job Instruction Training follows the PDCA cycle.

Step 1—Prepare the worker

The goal of this step is to put the worker at ease, establish some rapport with the employee rather than immediately jumping into the training. Find out what the trainee already knows about the job, if anything, and have a very informal discussion with them to put them at ease. At this initial discussion, it might be wise to explain to the employee that they will not pick up everything right away, and possibly give an example of a success story employee—someone who struggled through the training initially but passed and is thriving on the production floor. The job of the trainer is to reinforce that putting forth best effort during the training will lead to success, both in passing any training test required, but also in performing the job on the shop floor.

Steps 2 and 3—Overview

For steps 2 and 3, where the trainer presents the operation and then the trainee tries out the operation, each step should be done three times. The purpose of the training is to set the employee up for success in performing the job by giving them the needed knowledge and skills necessary to perform the task. If the training is being done in a small group, the trainer need only go through this process one time for the entire group, and then allow each employee to try out performance. Taking the time to walk through the process three times, and then have each employee perform the process four times takes time. Companies such as Toyota claim that coaching standardized work is the lengthiest step in a conversion to TPS (Huntzinger, 2006).

In the JI training, the trainer will be presenting the process to the employee, and then the employee will perform the process. In addition to performing the actual steps of the process, both the trainer and employee will speak aloud what they are doing, how they are doing it, and the reasons or key points for each step.

Having the process presented to the employee while explaining how to perform the task and why each task is important, and then having the employee do the same process has many benefits in helping the employee retain the information. The JI training is akin to the see–say–do multisensory learning strategy. Multisensory learning combines two or more senses; in this learning, you frequently will use the following four common sensory-related processes:

1. Visualizing—creating mental images of pictures and colors
2. Verbalizing—process of speaking out loud to activate the auditory channel and build auditory memory

3. Reciting—process of explaining information out loud in your own words, without looking at printed information

4. Developing muscle memory—a kinesthetic process that involves performing multiple repetitions of steps or actions until the actions become automatic (Wong, 2015).

The see–say–do learning strategy helps learning by

- Boosting memory by coding information in more than one way or with more than one sensory channel
- Creating stronger sensory paths into long-term memory so information is more clearly imprinted in your long-term memory
- Creates multiple ways for you to access and recall information at later times
- Adding motivation and interest to the learning process (Wong, 2015).

Step 2—How to Instruct

In preparing to present the operation to the trainee, line up the trainee so that they have the same view of the process as the trainer. Often, the trainee will need to be looking over the trainer's shoulder as the trainer performs the operation. Viewing the operation from the viewpoint that the trainee will be performing the process allows better retention, as they are watching the process from the angle in which they will also be doing the process. The point is to make sure that the employee has the exact viewpoint and can clearly see everything that you are describing to them.

In presenting the operation, the trainer will go through each part three times, so completing the "present the operation section" for a three-step process will look like this:

1. What
2. What, how
3. What, how, why

This JI process of making a sandwich would proceed with the trainer going over the following steps, performing each step and stating the following as they go.

First Pass, go through each Step in the process.

1. Select Bread
2. Cut Bread to Size
3. Cut Open Bread
4. Place Cheese(s)
5. Place Meat(s)

6. Place Vegetable(s)
7. Place Condiment(s)
8. Close Sandwich
9. Wrap Sandwich

For Second Pass, go through each step in the process, also going over the key points, or how the process is to be performed. So, for the second time through the process, as the trainer is performing each step, the trainer will state the step and key points, which are the numbered steps and letters underneath. For the third and final time through the process, the trainer will go through each step, each key point, and then add the reasons for each key point. To lessen redundancy, the reasons why are added under each key point.

1. Select Bread
    a. Bread type is listed on the ticket.
        i. Check the ticket to make sure that the proper bread is selected.
    b. Bread is labeled by type in Rack A.
        i. Central location for all bread enables quick retrieval of the item.
2. Cut Bread to Size
    a. Bread size listed on the ticket—options are 8 or 12 inches.
        i. Make sure that the proper size is selected.
    b. Line up the bread on ruler reading left to right.
        i. 0 on ruler begins on the left.
    c. Push down the bread gently with your left hand just before the cut line.
        i. This prevents the bread from sliding during cutting.
    d. Cut the bread straight down with a Bread Knife.
        i. Cutting straight down ensures that the sandwich is of accurate length.
3. Cut Open Bread
    a. Move the bread alignment from horizontal to vertical.
        i. This enables easier cutting of the bread.
    b. Move your left hand to the left end of the bread and push down gently.
        i. This keeps the hand out of knife cut area, keeps the bread from moving, and prevents the knife from cutting completely through the bread.

   c. Starting at the top right of the bread, angle knife slightly, cutting down until the entire piece of the bread has an opening.

      i. This opens the bread to allow for contents to be placed.

   d. Move the bread back to the horizontal position.

      i. To slide down the table and insert contents.

4. Place Cheese(s)

   a. Cheese(s) selection is listed on the ticket—options are American, Provolone, or Cheddar.

      i. Make sure that the proper cheese is selected.

   b. Grab a small stack (three to five pieces) with your left hand.

      i. Cheese sticks together, so it is easier to grab a smaller stack and peel off each piece in front of you, than to peel off each piece while the cheese is still in the bin.

   c. With the cheese in your left hand, peel individual slices with your right hand.

      i. Peeling one piece at a time while holding a small stack close to you is easier and more ergonomic than reaching into the bin for each piece.

   d. Place cheese circles evenly on the sandwich on the side of the bread closest to you—an 8-inch sandwich gets two pieces, whereas a 12-inch sandwich gets three pieces.

      i. That many pieces is the standard.

   e. Return extra cheese to the bin.

      i. So it can be used for another sandwich.

5. Place Meat(s)

   a. Meat(s) selection is listed on the ticket.

      i. Make sure that the proper meat is selected.

   b. Pick up the meat with two hands and pull back the paper on the bottom.

      i. Meat comes in premade packets, so the paper must be removed.

   c. Line up the left end of the meat with the left end of the sandwich.

      i. These should correspond in length, so lining each up at the same end will ensure that the meat fits properly into the sandwich.

   d. Place the meat on the sandwich on the side closest to you and remove the paper.

      i. After putting the meat on the sandwich, the top layer of the paper must be removed.

6. Place Vegetables
   a. Vegetable(s) selection is on the ticket.
       i. Make sure that the proper vegetable is selected.
   b. For an 8-inch sandwich—grab a handful of vegetable and starting at the left end of the sandwich, evenly disburse on the sandwich on the side closest to you.
       i. Starting at the left end gives a starting point. A handful gives enough for an 8-inch sandwich.
   c. For the 12-inch sandwich—grab an extra half-handful and evenly disburse on the remainder of the sandwich.
       i. Same as in 8-inch sandwich, but additional quantity is needed for a larger sandwich.
7. Place Condiments
   a. Condiment(s) selection is on the ticket.
       i. Make sure that the proper condiment is selected.
   b. Grabbing the condiment, turn it upside down as you are bringing over the sandwich.
       i. It saves time to begin turning the bottle over as you bring it over the sandwich.
   c. Beginning at the left end of the sandwich, squeeze the middle of the bottle over the sandwich on the side closest to you, then run left to right at a steady pace to make one line, then back to right to left to make a second line.
       i. Starting at left and making two lines ensures a consistent application of the condiment.
   d. Return the condiment with its right side up.
       i. The bottle will leak if left upside down.
8. Close Sandwich
   a. Roll the sandwich on its side, with the open side facing up.
       i. This gives access to the contents of the sandwich easily.
   b. Press the contents of the sandwich down toward the table with a sandwich knife.
       i. This compacts items into the sandwich.
9. Wrap Sandwich
   a. Pick up the sandwich with both hands, with hands over the open end of the sandwich.
       i. This prevents contents from falling out.
   b. Place the sandwich at the bottom and middle of the wrap paper.
       i. Must be at bottom for rolling.

    c. Roll the sandwich end over end until you reach the top end of the wrap paper.

       i. This wraps the sandwich completely.

    d. Fold the ends of the wrap paper toward the center.

       i. This secures loose paper.

    e. Tape the end of the paper by applying one piece of the tape that connects the two ends.

       i. This secures the paper so the sandwich doesn't fall out.

Going through the steps of the process three times before the trainees are asked to perform the process gives them necessary confidence and an initial understanding of the process. After the trainer has performed the steps three times, ask the trainee to try the process, reaffirming to them that they are still training, so now is the time to make mistakes. Putting the worker at ease initially and continuing to give them the latitude to make mistakes makes the trainee more comfortable with asking questions and truly understanding the process, as they know they are not expected to be experts yet.

Step 3—How to Instruct

During the first attempt at the process, have the trainee do the process without speaking and correct any mistakes that are made. Have the trainee do the process at least three more times, following the preceding format (first time have them state the step, second time the step and the key point (the "how"), and the final time, the step, key point, and reasons why). The multisensory learning not only adds interest and motivation in the learning process, but it also boosts memory and enables the trainee to call upon information at later times.

Step 4—How to Instruct

Some processes such as the sandwich-making example are more robust than others, and may require additional attempts by the trainee in order to master the steps and remember the reasons why. Take the time needed during the training, as the training card says, and continue until you know that the trainee knows.

After the trainees have completed the class training, place them with a "certified trainer," and have them begin performing the process on the line incrementally (starting with every third cycle, then every other, and finally every cycle). This will continue to build confidence in the trainees so that when they are on their own, they will have the necessary knowledge needed to perform the process and not only be empowered, but have lower stress. Once the trainees are on their own, the Job Instruction Card says to designate someone that they should go to for help. The trainee should be checked on frequently, and questions should be encouraged. Taper off extra coaching and close follow-up.

The workers on the floor perform the same process every day and are therefore the experts who know exactly how everything works, and how everything should look, feel, smell, etc. Thus, it is important to give any additional information to the employees that will enable them to have all possible knowledge of the process and inform them that they should immediately raise a hand (i.e., stop the line or get a team leader/supervisor) when something doesn't meet these specifications.

It is vital to explain each detail of the items to the employee, and identify the exact tools that are to be used. An example from the sandwich example could be that the cheese always comes in circles, so the employee should ask the question if the cheese they are using is now squares. When I worked at Toyota, we had a process that used two different-sized nuts. It was obviously important to use each nut for its intended purpose, and so when I first trained on the floor with an employee, the first thing they showed me was the difference between the two, where each was kept, where each was used in the process, and why it was important to use each for its specified use. Needless to say, I had all the information I needed to confidently perform the job and did not use the incorrect nut.

Even after employees have been trained, it may be necessary to return to the fundamentals to ensure that they properly understand the process. At the start of training camp in 1961, famed Green Bay Packers Coach Vince Lombardi began a tradition of starting from scratch, assuming that the players were blank slates who carried over no knowledge from the year before.... He began with the most elemental statement of all. "Gentlemen," he said, holding a pigskin in his right hand, "this is a football" (Maraniss, 1999). The purpose of this activity is to get everyone who performs the process aligned on the exact process (i.e., the standard), as there may be variations that some perform and others don't, or everyone performing the process may have not had adequate initial training, so it may be proper to perform a follow-up.

Standardize and Sustain are the two most difficult 5S elements. The standardization you achieved by following the PDCA format lays the foundation for Lean, but you cannot keep your foundation strong unless you have systems in place to sustain the standardization. The key to sustainment is to put systems in place to audit the implemented processes, and to coach the workers to audit. Put differently, you need accountability and visibility.

Hopefully you successfully laid the groundwork for each process by properly going through the PDCA process and coaching employees at each stage of the Process Implementation. If done correctly, you will not need to extensively train the employees on how and why to perform the process a certain way, or where items should be kept when not in use and in process.

The reasons that employees do not follow the standards put in place are:

1. Not properly trained
2. Trained but still unsure about what to do
3. New wrinkle in the process that they do not understand
4. Thought that their way is better
5. Obstacle keeping them from doing it
6. Motivation
7. No repercussions for doing it wrong

To sustain, you need to properly address all of these potential issues.

To resolve the first four potential issues, coaching the employees is crucial; also, as previously discussed, employees need the proper foundation to perform each process, so that when you follow up on a potential issue, you are asking questions and not questioning (what has been previously discussed as ensuring that a proper Plan (i.e., training and job knowledge are in place), is in place, before "Checking" on the issue. Basically, you can't "question" why someone performed the task a certain way if they have not had the proper training and given the proper understanding of the rationale behind each step. Because employees need coaching even after they have been trained, you will need to continue to coach employees through this phase to ensure that they know all the intricacies of the processes and are properly auditing the area, but note that coaching can eliminate many of the issues for why processes are done wrong and not sustained.

For items 6 and 7, at some point, coaching must cease and discipline must take over. I will not try to lay out the point for when exactly this occurs, but I will say that make sure the employee has had enough coaching before you resort to discipline. The first time an employee has an issue with a process should not be an automatic termination, but determine the root cause and coach the employee on the issue.

MFWC—MULTI-FUNCTION WORKER CHART

In the training section, and in other sections, we have discussed tracking the skill level of the employees and having new employees train with a "certified trainer." The MFWC is a tool that allows you to track this skill level. The MFWC also provides a means to set a training timetable for employees to increase their skill level for different processes and gives a visual of cross-training needs within the group.

To complete the MFWC, fill in the name of the line/group, the shift, and the date that the MFWC is being completed or updated. Not only does this accomplish filling in the vital information, but it also assists in auditing the MFWC, as it is important to know the date of the last update. The columns above the blank circles are for the process name. These blanks can also list the job titles, but if job titles are listed, there must be some document connecting the standardized work sheets with each job title (e.g., each title should have a list of corresponding standardized work sheets that they are to know). If you have never seen an MFWC, the rest likely does not make much sense at first glance. The remaining blanks allow you to make the skill level and training plans for each employee visual.

As shown in Figure 11.3, which illustrates an example of an MFWC, the MFWC should list all employees in the group as well as their skill level at each process. Each employee has a skill level for each process, even if that skill level is 0, and this skill level is identified by the shading of the circle (shading explained in detail below). Supervisors are responsible for determining the initial skill level as well as updating the MFWC as to any increased or new skill level; supervisors should have the best idea of each person's skill level, as they are making daily labor assignments and are the frontline person notified when something goes wrong.

Once each employee and process is listed on the MFWC, determine what skill level each employee has for each process and color in the circle accordingly. The MFWC in Figure 11.3 shows the skill level key in the bottom right corner; the following is an explanation of each of the four skill levels:

- 25%, or 1/4 of the circle filled in, means that the employee has been trained on the process by a Job Instruction (JI) certified trainer.
- 50%, or 1/2 the circle filled in, means that the employee performs to quality, but not in the required takt time (the allotted time to complete the task).

Multi-function worker training plan

Line/area:		Process name													Remarks			
Shift:																Capabilities			Manpower needs
Date:																Jan.	Jun.	Dec.	Performance needs
Name: Nu	Doug Peters	⊕⊕⊕⊕⊕⊕⊕⊕⊕⊕⊕																	
	Due date:																			
Name:		⊕⊕⊕⊕⊕⊕⊕⊕⊕⊕⊕																	
	Due date:																			
Name:		⊕⊕⊕⊕⊕⊕⊕⊕⊕⊕⊕																	
	Due date:																			
Name:		⊕⊕⊕⊕⊕⊕⊕⊕⊕⊕⊕																	
	Due date:																			
Name:		⊕⊕⊕⊕⊕⊕⊕⊕⊕⊕⊕																	
	Due date:																			
Name:		⊕⊕⊕⊕⊕⊕⊕⊕⊕⊕⊕																	
	Due date:																			
Name:		⊕⊕⊕⊕⊕⊕⊕⊕⊕⊕⊕																	
	Due date:																			
Name:		⊕⊕⊕⊕⊕⊕⊕⊕⊕⊕⊕																	
	Due date:																			
Result of training	Beginning of year														● 100% JI certified				
	Middle of year														◕ 75% Performs to quality in required time				
	End of year														◑ 50% Performs to quality, not in required time				
Remarks	Job needs														⊕ 25% Received JI training on process				

FIGURE 11.3
Blank Multi-Function Worker Chart.

- 75%, or 3/4 of the circle filled in, means that the employee performs to quality in the required takt time.
- If the employee is 100%, meaning that the entire circle is filled in, they are qualified to teach the process, and are JI-certified, a "certified trainer."

Having each employee's level at each process documented provides a visual to assist in the following:

1. Setting training schedules for employees—After filling in everyone's skill level for each process, you may notice that only one person is qualified to perform a process. On the MFWC, you can designate when someone should be at a certain level (e.g., John to be at 75% by June 4). Develop a training plan and post it with the MFWC so that it remains visible to everyone.
   a. When training an employee on a process, follow the JI Training guidelines.

2. Allows for quick readjustment of skill—In the scramble of trying to staff each position in the morning, this sheet can be used to place employees on positions that they are qualified for when the normal set up is hampered by someone not being at work (e.g., vacation, sick, leave, call off).

3. Motivation to employees—If employees see that they are only 50% or 75% on a process, it will motivate them (if they are properly empowered and involved), and they will ask what they need to do in order to increase their skill level.

The skill level for each employee in the group will drive how the rest of the MFWC is used. To identify gaps in training, set a goal as to how many employees are needed to be at 75% and 100% for each process. This goal will then dictate your cross-training plans.

- The "Due Date" is the date that the employee will reach the next skill level (i.e., if the employee is at 50% and the due date is August 12, then a plan should be in place for the employee to reach 75% by August 12).
- Capabilities are the number of processes that the employee will be 75% or 100% at the designated times; an additional barometer to drive cross-training.
- The Result of Training tracks the number of employees at 75% or 100% at a certain process for different times of the year; establish a goal for how many employees should be at this level for each process, and work toward that goal with cross-training. The tracking aspect allows you to see any manpower gaps, and work to close these gaps by setting goals for cross-training, with date targets for each employee at each process.

Having everything visual on the MFWC creates accountability and starts the planning process for training.

As with everything else that has been implemented, the MFWC should also be audited. In auditing the MFWC:

1. Make sure that the employee roster is up to date
2. Make sure that the employee skills are up to date
3. Ensure that employee training plans are on pace
4. Identify cross-training needs—make sure training plans are being implemented

The MFWC does not need to be audited daily. Set a standard audit schedule for the MFWC that will sustain the system, as well as someone, likely a supervisor, who is responsible not only for maintaining the skill set and tracking the training, but also for making any necessary changes as needed (e.g., changing employee roster, updating skill for each process).

REFERENCES

Huntzinger, Jim, *Why Standard Work Is not Standard: Training within Industry Provides an Answer*, Target Volume 22, No. 4, p. 11, 2006.

Liker, Jeffrey and David Meier, *Toyota Talent, Developing Your People the Toyota Way*, p. 88, McGraw-Hill, New York, 2007.

Maraniss, David, *When Pride Still Mattered: A Life of Vince Lombardi*, Simon and Schuster, New York, 1999.

Mayfield, Jacqueline and Milton Mayfield, *Leader Communication Strategies; Critical Paths to Improving Employee Commitment*, American Business Review, p. 90, June 2002.

Salvendy, Gavriel, *Handbook of Human Factors and Ergonomics*, Wiley publisher, p. 413, 2012.

Williams, Scott, *Effective Empowerment: Boosting Confidence and Removing Obstacles*, Leader Letter, Wright State College of Business, 2015.

Wong, Linda, *Essential Study Skills*, Wadsworth Cengage Learning, Boston, p. 13, 2015.

Conclusion

Attempting to implement or practice Lean methods without setting the proper foundation would be unsuccessful. To properly utilize Lean, the standardized work and 5S foundation must be strong, as it is the base on which all other Lean elements stand. By following PDCA, you are now ready to implement new processes in your work area. Implementing these processes will provide your company with stability by establishing the foundation for Lean through standardized work and visual controls. However large the project you are about to undertake, following the process prescribed in this book will make the implementation smooth, and the end product better.

Index

Page numbers with f and n refer to figures and footnotes, respectively